T0300247

CLOSED-LOOP SUPPLY CHAINS

New Developments to Improve the Sustainability of Business Practices

SUPPLY CHAIN INTEGRATION
Modeling, Optimization, and Applications

Sameer Kumar, Series Advisor
University of St. Thomas, Minneapolis, MN

Closed-Loop Supply Chains: New Developments to Improve the Sustainability of Business Practices
Mark E. Ferguson and Gilvan C. Souza
ISBN: 978-1-4200-9525-8

Connective Technologies in the Supply Chain
Sameer Kumar
ISBN: 978-1-4200-4349-5

Financial Models and Tools for Managing Lean Manufacturing
Sameer Kumar and David Meade
ISBN: 978-0-8493-9185-9

Supply Chain Cost Control Using Activity-Based Management
Sameer Kumar and Matthew Zander
ISBN: 978-0-8493-8215-4

CLOSED-LOOP SUPPLY CHAINS

New Developments to Improve the Sustainability of Business Practices

Mark E. Ferguson and Gilvan C. Souza

CRC Press
Taylor & Francis Group
Boca Raton   London   New York

CRC Press is an imprint of the
Taylor & Francis Group, an **informa** business

Auerbach Publications
Taylor & Francis Group
6000 Broken Sound Parkway NW, Suite 300
Boca Raton, FL 33487-2742

International Standard Book Number: 978-1-4200-9525-8 (Hardback)

Library of Congress Cataloging-in-Publication Data

Closed-loop supply chains : new developments to improve the sustainability of business
    practices / editors, Mark E. Ferguson, Gilvan C. Souza.
      p. cm. -- (Supply chain integration series)
    Includes bibliographical references and index.
    ISBN 978-1-4200-9525-8 (hardcover : alk. paper)
    1. Business logistics 2. Business logistics--Environmental aspects. 3.
Remanufacturing. I. Ferguson, Mark, 1969- II. Souza, Gilvan C.

HD38.5.C55 2010
658.5--dc22
                                                        2010005524

Visit the Taylor & Francis Web site at
http://www.taylorandfrancis.com

and the Auerbach Web site at
http://www.auerbach-publications.com

Contents

Preface

Closed-loop supply chains are supply chains where, in addition to the typical forward flow of materials from suppliers to end customers, there are flows of products back (post consumer touch or use) to manufacturers. Examples include product returns flowing back from retailers to the original equipment manufacturers (OEMs), used products (with some remaining useful life) that are traded in for a discount on the purchase price of a new product, end-of-lease returns, and end-of-life products that are returned for disposal or recycling. Interest in the management of closed-loop supply chains has increased noticeably in the last ten years. Drivers of this increased interest include the substantial increase in the price of raw materials, the increase in consumer product returns (driven in part by the design of increasingly complex products), an increase in the awareness at the executive level of a firm's environmental footprint, pressure from customers and nongovernmental organizations to be better environmental stewards, and current and pending legislation requiring end-producer responsibility for its products at the end of their life. The increase in interest of this topic among academics is demonstrated by the creation of the College of Sustainable Operations inside the Production and Operations Management Society (POMS), a department exclusively dedicated to this topic in the *POM Journal* (and entirely separate from the supply chain management department), and the annual workshop of researchers in this field that has grown in size and interest over the last nine years.

The aim of this book is to provide both researchers and practitioners a concise and readable summary of the latest research in the closed-loop supply chain field, particularly when there is remanufacturing involved. In addition to current research topics, we provide examples of industries that have implemented profitable product recovery and remanufacturing operations. From these examples, we highlight common practices to provide guidance to firms that are not currently active in the secondary market for their products. The focus throughout this book is on business practices that are environmentally friendly and profitable. Thus, it is not our intention to make societal judgments on a particular business practice but rather to demonstrate the potential of increased profitability obtained from firms that take a proactive rather than reactive approach to current and pending environmental regulations and pressures.

This book is divided into four parts. Part I looks at the strategic decisions facing a firm with regard to the secondary market for its products, including the impact of environmental regulation. Part II looks at the tactical decisions assuming a firm has made the decision to remanufacture/refurbish in-house. Part III summarizes some key characteristics of different industries where remanufacturing is common and provides detailed case studies of companies running profitable reuse/remanufacture/recycling operations. Finally, Part IV addresses the need for expanding the research in this area beyond operations management to other disciplines in the business school and provides some future research directions.

The focus of Part I on strategic issues is on decisions that are typically made at the upper levels of management of OEMs. Examples of some strategic questions facing firms of durable and semi-durable products include the following:

- Should the firm interfere in the secondary market of its products?
- Should the firm offer a take-back or trade-in program to recover its products at the customer's end of use?
- If returned products are sold by the firm, should they be sold through the same channels as the firm's new products?
- If the firm chooses to recycle, refurbish, or remanufacture, should it be done in-house or outsourced?
- Should product design decisions be influenced by the end-of-use decision?

In Chapter 2, the focus is on an OEM's decision to participate (either actively or passively) in the secondary market of its products. Several opportunity costs are discussed here that should be factored into this decision. Some of these opportunity costs, such as the cost of the remanufactured products cannibalizing the sales of the OEM's new products, factor against the decision to remanufacture. Other opportunity costs, such as the opportunity for third-party entrants, support the OEM's decision to remanufacture. In Chapter 3, the authors categorize the latest environmental legislation around the world that relates to the OEM's responsibility of its products at the end of life. They also include a summary of what the academic research has to say on the effectiveness of the various proposed and enacted forms of this legislation to the various stakeholders: policy makers, firms, and the environment. Chapter 4 provides some general guidelines, as well as some case studies and examples, of design principles for closing the loop. Guidelines include product line architecture guidelines (e.g., using modular designs and using classic designs to avoid "fashion" obsolescence), product maintenance guidelines (to increase durability and serviceability), product standardization guidelines (to avoid unnecessary proliferation), and guidelines on the use of hazardous materials. In addition, there is a detailed discussion on specific hardware design guidelines, such as ease of inspection and sorting, disassembly, cleaning, reassembly, use of reusable components, and design for recycling.

In Part II, the focus switches to more tactical issues where the assumption is made that a firm has already decided to remanufacture and thus desires to do so in the most profitable manner possible. Examples of tactical questions facing firms that decide to remanufacture in-house are the following:

- What is the most efficient collection network to recover used cores?
- What should be done with products that are taken back? Should they be landfilled, incinerated, recycled, harvested for parts, sold as-is, refurbished, or remanufactured? (This is referred to as the disposition decision.)
- What is the value of pre-sorting the returned cores into different quality grades based on the amount of effort or expense to remanufacture? How many different quality grades are needed?
- How do you create a production plan for a remanufacturing operation? How is it different from a production plan for making new products?
- How should a firm market remanufactured products?

In Chapter 5, the focus is on designing the reverse logistics network for collection, processing, and remanufacturing of used products, as well as remarketing remanufactured products. The analysis includes channel structure (collection directly from consumers, or through third parties such as retailers); drop-off versus pick-up collection strategies; the use of financial incentives to improve collection rates; and the location of collection points, consolidation points, and remanufacturing facilities. In Chapter 6, three interconnected tactical decisions are discussed: product acquisition, grading, and disposition. Product acquisition refers to the process of acquiring used products (returns), which may come naturally (e.g., end-of-lease products), may be mandated by regulation, or may be proactively purchased by the firm. In some cases, the purchase price has a direct impact on the quality of acquired returns. Regardless of a proactive or reactive acquisition strategy, the firm must grade returns into different categories, according to their quality, which is correlated to the amount of labor and materials necessary to remanufacture the returns. Finally, after grading, the firm must make a disposition decision for each return, according to its quality category, expected demand, and revenue opportunities for different reuse options. As an example, the firm may decide that the worst-quality returns are to be recycled for materials recovery, the second worst category of returns should be used for harvesting spare parts, and the firm should remanufacture the remainder as long as there is demand. In Chapter 7, two specific production-planning methodologies are proposed to aid a firm in making disposition decisions, especially remanufacturing. It is assumed that the firm has a grading operation in place, and the firm has forecasts for returns and remanufactured products over a planning horizon. One methodology discussed in Chapter 7 uses optimization techniques in an environment where remanufacturing capacity is limited, whereas the other methodology is based on MRP logic and is best suited for environments with fewer capacity constraints. Finally, Chapter 8 provides an analysis of the market for remanufactured products, including the price differentials between remanufactured and new products observed empirically, the impact of seller reputation and warranties on demand for remanufactured products, and consumer (post-purchase) satisfaction with remanufactured products. The findings from Chapter 8 are based on a large-scale dataset regarding online purchase transactions of both new and remanufactured products across different product categories. Among other findings, the authors emphasize the critical importance of warranties and seller reputation on consumer willingness-to-pay for remanufactured products—even more critical than for corresponding new products.

The focus of Part III is on describing actual reuse/remanufacture/recycling practices in a wide variety of industries. Some of the industries have been described and studied before (such as the summaries of the retreaded tires, single-use cameras, toner cartridges in Chapter 9), so the chapter serves as an update on these industries. The practices of other industries such as the movie picture industry (Chapter 10) and health care, particularly hospitals (Chapter 11), have not received much attention previously. In addition, Chapter 9 identifies common characteristics across a broad sampling of industries that make remanufacturing more or less attractive.

Finally, Part IV focuses on summarizing related research in other fields and identifying future research opportunities in closed-loop supply chains. The outline of the book is as follows:

Chapter 1: A Commentary on Closed-Loop Supply Chains (Mark Ferguson and Gilvan C. Souza)

Part I: Strategic Considerations
Chapter 2: Strategic Issues in Closed-Loop Supply Chains with Remanufacturing (Mark Ferguson)
Chapter 3: Environmental Legislation on Product Take-Back and Recovery (Atalay Atasu and Luk N. Van Wassenhove)
Chapter 4: Product Design Issues (Bert Bras)

Part II: Tactical Considerations
Chapter 5: Designing the Reverse Logistics Network (Necati Aras, Tamer Boyacı, and Vedat Verter)
Chapter 6: Product Acquisition, Grading, and Disposition Decisions (Moritz Fleischmann, Michael R. Galbreth, and George Tagaras)
Chapter 7: Production Planning and Control for Remanufacturing (Gilvan C. Souza)
Chapter 8: The Market for Remanufactured Products: Empirical Findings (Ravi Subramanian)

Acknowledgments

Mark Ferguson's special thanks:

I would like to thank my coauthors, colleagues, and students who have helped open my eyes to the need for more sustainable business practices, and my wife, Kathy, and daughters, Grace and Tate, for their love, encouragement, and support.

Gil Souza's special thanks:

I would like to thank the participants and organizers of the workshop on closed-loop supply chains over the years—several are coauthors on many projects, many are close friends, and my interaction with them shaped my interest and understanding of the subject over the years. I would also like to thank my friends and family for encouragement and support over the years.

Editors

Mark Ferguson is the Steven A. Denning Professor of Technology and Management and the John and Wendi Wells Associate Professor of Operations Management in the College of Management at Georgia Institute of Technology, Atlanta. He received his PhD in business administration, with a focus in operations management from Duke University in 2001. He holds a BS in mechanical engineering from Virginia Polytechnic Institute and State University, Blacksburg, and an MS in industrial engineering from Georgia Institute of Technology. Currently, he serves as the faculty director of the technology and management program at Georgia Institute of Technology—a joint program between the colleges of management and engineering. His research interests involve many areas of supply chain management including supply chain design for sustainable operations, contracts that improve overall supply chain efficiency, pricing and revenue management, and the management of perishable products. Dr. Ferguson serves as the coordinator for a focused research area on dynamic pricing and revenue management. Two of his papers have won best paper awards from the Production and Operations Management Society and several of his research projects have been funded by the National Science Foundation. Prior to joining Georgia Institute of Technology, he had five years of experience as a manufacturing engineer and inventory manager with IBM.

Gilvan "Gil" C. Souza received his BS in aeronautical engineering from Instituto Tecnológico de Aeronáutica (ITA), Brazil; his MBA from Clemson University, South Carolina; and his PhD in operations management from the University of North Carolina. Before entering academia, he was a product development engineer at Volkswagen in Brazil for several years. He is currently an associate professor of operations management at the Kelley School of Business, Indiana University, Bloomington. Prior to this, he was on the faculty at the Smith School of Business, University of Maryland, College Park. Dr. Souza is the author or coauthor of several research papers published in *California Management Review*; the *European Journal of Operational Research*; *Management Science*; *Manufacturing and Service Operations Management*; and *Production and Operations Management*. His current research interests lie in supply chain management, including production planning, remanufacturing, and sustainable operations. He was the recipient of the Wickham Skinner Early-Career Research Accomplishments Award from the Production and Operations Management Society (POMS) in 2004, and the Skinner best paper award from POMS in 2008.

Contributors

Vishal Agrawal
College of Management
Georgia Institute of Technology
Atlanta, Georgia

Necati Aras
Department of Industrial Engineering
Boğaziçi University
Istanbul, Turkey

Atalay Atasu
College of Management
Georgia Institute of Technology
Atlanta, Georgia

Martin Beaulieu
HEC Montréal
Montréal, Quebec, Canada

Tamer Boyacı
Desautels Faculty of Management
McGill University
Montréal, Quebec, Canada

Bert Bras
George W. Woodruff School of
    Mechanical Engineering
Georgia Institute of Technology
Atlanta, Georgia

Charles J. Corbett
Anderson School of Management
University of California, Los Angeles
Los Angeles, California

Mark Ferguson
College of Management
Georgia Institute of Technology
Atlanta, Georgia

Moritz Fleischmann
University of Mannheim
Business School
Mannheim, Germany

Michael R. Galbreth
Moore School of Business
University of South Carolina
Columbia, South Carolina

Robert D. Klassen
Richard Ivey School of Business
University of Western Ontario
London, Ontario, Canada

Sylvain Landry
HEC Montréal
Montréal, Quebec, Canada

Gilvan C. Souza
Kelley School of Business
Indiana University
Bloomington, Indiana

Ravi Subramanian
College of Management
Georgia Institute of Technology
Atlanta, Georgia

George Tagaras
Department of Mechanical
  Engineering
Aristotle University of Thessaloniki
Thessaloniki, Greece

L. Beril Toktay
College of Management
Georgia Institute of Technology
Atlanta, Georgia

Rajesh K. Tyagi
HEC Montréal
Montréal, Quebec, Canada

Stephan Vachon
HEC Montréal
Montréal, Quebec, Canada

Luk N. Van Wassenhove
Social Innovation Center
INSEAD
Fontainebleau, France

Vedat Verter
Desautels Faculty of Management
McGill University
Montréal, Quebec, Canada

Chapter 1

Commentary on Closed-Loop Supply Chains

Mark Ferguson and Gilvan C. Souza

Content

The sustainability movement has gained significant momentum over the last few years as both consumers and corporate managers begin to realize the impact of unsustainable environmental practices on their current and future quality of living standards and profits. The most immediate and direct impact of environmental issues for most people has been the recent dramatic increase in the cost for fossil fuels and raw materials. Not surprisingly, issues regarding energy usage, access to clean water, carbon dioxide emissions, and climate change have received the vast majority of the attention in the popular press. Each of these areas are indeed critically important, but there is at least one additional issue facing countries across the world whose long-term effects may be just as critical and potentially life changing as the ones discussed above. This less-publicized issue is the increasing rate of land-filling with manufactured products made of depletable raw materials and resources. Simply put, the current business practice of extracting raw materials from the earth, manufacturing them into products, and then disposing of the products into landfills or incinerators after a short period of use is not sustainable. For example, depending on estimates about current recycling rates, we could run out of zinc by 2037, run out of indium and hafnium (used in computer chips) by 2017, and run out of terbium (used in fluorescent lights) by 2012 (Cohen 2007).

In addition, the availability of land available for product disposal will be used up, leading to a significant reduction in the fortunes of pure product-based companies and a lower standard of living for consumers around the world. The numbers demonstrating the problem are hard to fathom. Each household in the United Kingdom generates approximately 1 ton of waste each year. Even worse, for every ton of products we buy, 10 tons of resources are used to produce them.* In the United States, each person generates approximately 4.6 pounds of waste per day for a cumulative total of 251 tons of solid waste that were either incinerated or sent to landfills in the year 2006. Of these 251 tons, 16 percent were categorized as durable goods. The disposal of durable goods is particularly troublesome because they are often manufactured using material from nonrenewable resources. The only sustainable business practice for producing durable goods is to reuse or recover the nonrenewable materials they are made of. Unfortunately, of the 40.2 million tons by weight of durable goods sold in the United States in 2006, only 18.5 percent of the material used in their production has been, or is expected to be, recovered.[†] Most manufacturers of durable goods recognize this fact and are starting to devise strategies for their long-term survival, strategies that involve dramatic changes in the way they have historically viewed their supply chains.

As demonstrated above, recycling of raw materials is clearly one important sustainability activity; however, there are other practices, such as remanufacturing, that may have an even higher positive environmental impact in some industries.[‡] We now define closed-loop supply chains and briefly define and discuss other disposition decisions.

Closed-loop supply chains are supply chains where, in addition to the typical "forward" flow of materials from suppliers all the way to end customers, there are flows of products back (post-consumer touch or use) to manufacturers. An example of closed-loop supply chain, adapted from Ferguson et al. (2009), is shown in Figure 1.1. Pitney Bowes (PB) is an original equipment manufacturer (OEM) headquartered in Stamford, CT, that manufactures large-scale mailing equipment. Functions performed by these machines include matching customized documents to envelopes, postage printing based on weight, and sorting mail by zip code (due to contracts with the U.S. postal service, sorting mail is a source of significant savings for companies that mail large

---

* http://www.wasteonline.org.uk/resources/InformationSheets/HistoryofWaste.htm
† EPA-530-F-07-030, November 2007, www.epa.gov/osw
‡ For a good overview of the process of remanufacturing, we refer to the research performed by Nabil Nasr and his associates at the The Golisano Institute of Sustainability at Rochester Institute of Technology (www.sustainability.rit.edu).

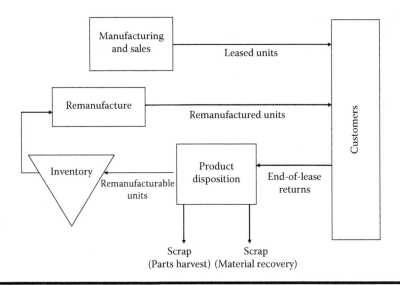

Figure 1.1   Closed-loop supply chain for Pitney Bowes.

quantities of documents). PB mainly leases its equipment; on average 90 percent of PB's revenues are derived from leasing. A typical leasing agreement is for four years. At the time of the leasing contract renewal, customers may opt for equipment of a newer technological generation (if available). In that case, customers return their end-of-lease products to PB. All used equipment is tested and sorted. A disposition decision is then made for each individual machine; options include recycling (raw material recovery), parts harvesting (to recover parts for use in service contracts), or remanufacturing, which restores the used product to a common standard. Remanufactured products are sold at a discount relative to the new product's list price.

There are essentially three types of returns in closed-loop supply chains:

- *Consumer returns*: These returns originate from retailers that set "no questions asked" returns policies. For example, about 5–6 percent of newly sold printers are eventually returned, for various reasons (defects is not typically one of them), within the grace period of typical retailers—typically 15–30 days (Ferguson et al. 2006). Thus, consumer returns are technologically current, and have only been lightly used by the customer.
- *End-of-use returns*: These products have been used to a significant extent by the customer and consequently are of an older technological generation. Many, if not most, however, are still fully functional. Examples include cell phones (the average customer upgrades cell phones every 18 months in the Western countries*), PB's end-of-lease equipment as described above, and trade-ins.

---

* http://secret-life.org/index.php

■ *End-of-life returns*: These returns reached the end of their useful life; appropriate disposition decisions for these products include energy and materials recovery. Examples include very old electronic equipment that are nonfunctional or very expensive to repair, worn-out tires, and old carpet. For example, it is estimated that complete carpet recycling can recover $750 million in materials annually in the United States (Realff et al. 2004).

Returns are referred to as cores in certain remanufacturing industries. Disposition decisions for product returns include

■ *Landfilling*: This option is illegal for some products in some jurisdictions. For example, most states in the United States ban the landfilling of hazardous waste; electronic equipment is considered hazardous in states such as California, Maine, Massachusetts, and Minnesota (U.S. GAO 2005).

■ *Incineration*: Incineration helps to reduce the amount of solid waste going to landfills. For example, incineration can reduce the volume of solid waste by as much as 95 percent. Incineration can and is frequently used for energy recovery (energy from waste). It is thus an important option in countries and municipalities that have limited areas for landfilling, such as those in Europe. For example, although estimates vary somewhat, Denmark incinerates 58 percent of its municipal solid waste toward energy recovery, compared to about 11 percent for the United States (Knox 2005). The major drawbacks of incineration relate to emissions and pollution. For example, it is estimated that incinerators emit 446 kg/year of mercury in Canada (Knox 2005). In the United States, the Environmental Protection Agency (EPA) regulates incinerator emissions. Although incineration is the most proven technology for converting waste into energy, there are other technologies including gasification, pyrolysis, and plasma conversion (Knox 2005). Incineration is thus one step better than landfilling; however, it does not close the loop, as recycling and remanufacturing (next) do.

■ *Recycling*: This option implies materials recovery. This disposition option is attractive for returns with limited or no functionality remaining, and whose materials can be economically separated in an environmentally friendly manner. End-of-life returns, such as very old electronic equipment, are frequently recycled; in that case the product is shredded for posterior material separation (e.g., plastic, steel, aluminum, precious metals), and recycling of each material type. Recycling may be optimal, from an environmental perspective, for end-of-use returns such as older appliances; this is because newer appliances consume much less energy (Quarigasi Frota Neto et al. 2007). Even consumer returns, which are fully functional and technologically current, may face recycling, due to negative profitability associated with light refurbishing and remarketing of the product; an example is low-end printers at Hewlett-Packard (Guide et al. 2006). Recycling can be mandated by legislation; an example is the European Directive on Waste of Electrical and Electronic Equipment (WEEE), which mandates 65 percent recycling of collected electrical and electronic used products (by weight).

■ *Parts harvesting*: This option implies recovering selected parts from returns for use in service contracts (spare parts). This is a common practice in firms such as PB, Hewlett-Packard, and IBM. For example, it is estimated that IBM saves as much as 80 percent per part (destined to fulfill service contracts with customers) by dismantling returns compared to sourcing a new part from a supplier (Fleischmann et al. 2002).

■ *Resale (as-is)*: This option may be attractive if there exists an active secondary market for used equipment. For example, IBM sells some of their used IT equipment recovered from end of lease to certified brokers, who may refurbish or remarket them.

■ *Internal reuse*: This option implies light or no refurbishing: containers are an example.

■ *Remanufacturing or refurbishing*: This is a value-added operation, and has the potential for higher profitability among disposition decisions. Hauser and Lund (2003) define remanufacturing as an extensive process of restoring used products to "like-new" condition, including disassembly, cleaning, repairing and replacing parts, and reassembly. Refurbishing can be defined as "light" remanufacturing, and it typically involves little disassembly. We use the terms remanufacturing and refurbishing interchangeably in this book, except when explicitly noted.

We focus our attention in this book on closed-loop supply chains that include some level of remanufacturing or refurbishing, as remanufacturing is a value-added operation providing economic benefits and environmental benefits due to the extension of the product's useful life and reduced energy and material consumption (Hauser and Lund 2003). We do not focus on other environmental management practices (e.g., pollution prevention, reduction of energy consumption, and other sustainability practices) although improvements in product and material reuse typically improves these other dimensions as well.

References

Cohen, D. 2007. Earth's natural wealth: An audit. *New Scientist* 2605, 34–41.

Ferguson, M., V. Daniel Guide Jr., and G. C. Souza. 2006. Supply chain coordination for false failure returns. *Manufacturing & Service Operations Management* 8, 376–393.

Ferguson, M., V. Daniel Guide Jr., E. Koca, and G. C. Souza. 2009. The value of quality grading in remanufacturing. *Production and Operations Management* 18, 3.

Fleischmann, M., J. van Nunen, and B. Grave. 2002. Integrating Closed-Loop Supply Chains and Spare Parts Management at IBM. ERIM Report Series Reference No. ERS-2002-107-LIS. Available at http://ssrn.com/abstract=371054.

Guide Jr., V.D., G. Souza, L. N. Van Wassenhove, and J.D. Blackburn. 2006. Time value of commercial product returns. *Management Science* 52, 1200–1214.

Hauser, W. and R. Lund. 2003. *The Remanufacturing Industry: Anatomy of a Giant.* Boston, MA: Boston University.

Knox, A. 2005. An Overview of Incineration and EFW Technology as Applied to the Management of Municipal Solid Waste (MSW). Report available at http://www.oneia.ca/files/EFW%20-%20Knox.pdf.

Quarigasi Frota Neto, J., G. Walther, J. Bloemhof-Ruwaard, Nunen, J. A. E. E., and T. van Spengler. 2007. From Closed-Loop to Sustainable Supply Chains: The WEEE Case. *Working Paper.* Erasmus University, Rotterdam. Available at http://hdl.handle.net/1765/10176.

Realff, M., J. Ammons, and D. Newton. 2004. Robust reverse production system design for carpet recycling. *IIE Transactions* 36, 767–776.

U.S. GAO. 2005. U.S. GAO (Government Accountability Office) Report No. 06–47. November 2005.

STRATEGIC
CONSIDERATIONS

I

STRATEGIC CONSIDERATIONS

Chapter 2

Strategic Issues in Closed-Loop Supply Chains with Remanufacturing

Mark Ferguson

Contents

2.1 Introduction

Many statistics point to the need to find solutions for reducing waste. For example, in 2006, municipal solid waste amounted to more than 251 million tons (U.S. EPA 2007). To reduce waste, the U.S. Environmental Protection Agency recommends adopting a reduce–reuse–recycle hierarchy and resorting to combustion or landfilling only as a last resort (U.S. EPA 2008). Despite this recommendation, 67.5 percent of the municipal waste went directly to landfills or incineration facilities in 2006 (U.S. EPA 2007). Thus, it is encouraging that there is a market for remanufactured products in the United States. According to Hauser and Lund (2008), there are at least 2000, possibly up to 9000, firms in the United States that claim themselves as remanufacturers; if refurbishing is also included as being remanufacturing, these numbers will be larger. Examples of remanufactured products include automotive parts, cranes and forklifts, furniture, medical equipment, pallets, personal computers, photocopiers, telephones, televisions, tires, and toner cartridges, among others. These products are put on the market by the original equipment manufacturer (OEM) or independent remanufacturers. Given the size and growing importance of the remanufacturing market, there is a growing interest in the academic research community to further understand and explore this topic.

The stream of research on this topic goes under names like reverse logistics, green supply chains, and closed-loop supply chains. Until recently, the majority of research in this area has assumed that firms are actively involved in remanufacturing their own products and has thus focused on improving the efficiency of the processes needed to do so. Examples of these tactical, or operational, decisions include how to structure the reverse logistics network in an efficient manner, how and when returned units should be graded and processed, and what type of processing should be performed (disassemble to harvest parts and then build to order, remanufacture to stock, etc.).

The actual process of remanufacturing is almost always less expensive than producing a brand new unit of the product (at least on the margin) because many parts and components can be reused, thus avoiding the need to procure them from suppliers. In addition, by remanufacturing their used products, firms extend the products' life cycles, which helps keep them out of landfills. This practice should, in turn, improve the environmental perception of the company and help avoid negative publicity by environmental groups along with potential costly environmental legislation imposed on their industry. Indeed, there are many potential financial benefits to extending product life cycles besides the pure profit margin obtained by selling the remanufactured product. Despite all of these benefits from remanufacturing, as mentioned earlier, most firms continue to either ignore or, in some cases, actively try to deter any remanufacturing and reuse of their products. There are very few industries where all of the major companies in that industry participate in remanufacturing or product take-back initiatives at the same level of effort. What is more common is to find an industry where one company strongly embraces remanufacturing while a very similar-looking competitor to that company completely ignores it. From a management perspective, such situations are puzzling. If it is profitable for one company to be actively involved in the secondary market then why does not its competitor also choose to participate? In this chapter, we focus on the strategic decisions facing a firm regarding the secondary market for its products. As we will see, however, it is nearly impossible to completely separate the strategic decisions from the tactical problems.

The Xerox Corporation demonstrated early on that remanufacturing can be a very lucrative prospect (Berko-Boateng et al. 1993). In 1991, they obtained savings of around $200 million by remanufacturing copiers returned at the expiration of their lease contracts. Kodak is one of the classic examples of an OEM that has created a fully integrated manufacturing–remanufacturing strategy around its reusable Funsaver camera line (Toktay et al. 2000). Caterpillar is shifting its strategy from solely manufacturing and selling construction equipment to a leasing and remanufacturing strategy (Gutowski et al. 2001). This allows Caterpillar to create a new market among contractors who cannot afford to buy a Caterpillar product outright, but, instead, lease one when needed. From this early success, Caterpillar established a remanufacturing division that markets both equipment and parts, even including parts from other manufacturers. In 2007, this division had over $2 billion in sales and was the fastest growing division out of all of Caterpillar's divisions. In the same year, the Global Asset Recovery Solutions division of IBM collected over one million units of used information technology (IT) equipment that was converted to billions of dollars in revenues on the second-hand equipment, parts, and materials markets.

Unfortunately, from an environmental standpoint, the companies mentioned here represent the exception to the rule: most OEMs do not choose to remanufacture their products. In many of the cases where an OEM does not remanufacture, however, the void is filled by third-party firms whose primary business is to remanufacture the products of the major OEMs within a given industry. From their database of over 2000 remanufacturing firms, Hauser and Lund (2008) found that only 6 percent were OEMs. Third-party remanufacturing firms are often small to medium in size, with typical revenues in the range of $500,000–5,000,000. In response to the entry of these third-party firms, some OEMs actively try to deter the secondary market for their products by lobbying for legislation against the use of remanufactured products or creating internal policies such as voided warranties or stiff relicensing fees (for the case of IT equipment). The major OEM printer manufacturers, for instance, do not offer refilled printer cartridges themselves and are famous for their efforts (voided warranties, legal challenges, etc.) against the third-party cartridge refiller industry (http://www.rechargermag.com/). An example of regulation deterring third-party remanufacturers occurs in the aircraft engine overhauling business, where only the original engine manufacturers are allowed to "reset the clock" to zero for an overhauled engine. What is clear is that the practices of OEMs regarding remanufacturing are not consistent across industries, and even sometimes within an industry.

So why do some OEMs view remanufacturing as an opportunity while others appear to view it as a threat? We suspect that the question of how to position (or even to offer or not) a remanufactured product is not well understood by the majority of firms today. In the absence of analytical tools to help them, firms often develop rules of thumb such as to never price a remanufactured product more than *x* percent of the price of a new product. Bosch Tools division, for example, decides what product lines to remanufacture based upon the product's price and market share. If the market share is below a certain threshold and the new product price is above a given threshold, then the product is remanufactured, otherwise it is not (Valenta 2004). While such rules of thumb are common in practice due to the lack of guidance from scientific studies, the academic research has begun to shed some light on this important topic. In this chapter, we explore this topic and summarize some of the conclusions from recent works in this area.

2.2 Is Remanufacturing Profitable?

The bottom line for most companies that are struggling with the decision to remanufacture or not is whether or not offering a remanufactured product will increase profits. At first glance, this may seem like an easy question to answer. If the marginal cost to remanufacture a used unit is lower than the price the remanufactured product can be sold at, and the profit generated from this endeavor over a certain time period exceeds any fixed cost investment required to set up a remanufacturing operation and sales channel, then the firm should choose to do so.

The actual process of remanufacturing is almost always less expensive than producing a brand new unit of the product (at least on the margin) because many parts and components can be reused, thus avoiding the need to procure them from suppliers. In addition, by remanufacturing their used products, the firm extends the products' life cycles, which helps keep them out of landfills. This practice should, in turn, improve the environmental perception of the company and help avoid negative publicity by environmental groups along with potential costly environmental legislation imposed on their industry. Indeed, there are many potential financial benefits to extending product life cycles besides the pure profit margin obtained by selling the remanufactured product. Despite all of these benefits from remanufacturing, as mentioned earlier, most firms continue to either ignore or, in some cases, actively try to deter any remanufacturing and reuse of their products. Thus, we need to take a deeper look at this strategic decision to help understand some of the drivers behind it.

As previously mentioned, the marginal cost of remanufacturing is almost always lower than the marginal cost of producing a new unit, and there are many additional less tangible benefits to remanufacturing such as improving the firm's environmental reputation. So what keeps OEMs from remanufacturing their products? To begin with, many OEMs spend the majority of their time and resources focusing on their new product sales, and thus have simply not thought about remanufacturing as a viable business model. Even some OEMs that do take the time to seriously consider remanufacturing may be dissuaded because they do not feel they possess the infrastructure and expertise to collect the used units and remanufacture them in a profitable manner. This is a real concern for many—there has been a significant trend over the last 30 years for OEMs to outsource and offshore their manufacturing operations so that they can focus their resources on new product development, sales, and marketing. Without this original manufacturing expertise, it is often more difficult to set up a low-cost remanufacturing operation. So what about the contract manufacturers that make the new products for the OEM? Surely they must possess a thorough understanding of how to assemble the product and should thus be in the best position to remanufacture it at the lowest cost. Indeed, many OEMs who outsource the production of their new products (a list that includes almost all the major OEMs in the IT and electronics industries) to contract manufacturers also do so if they choose to remanufacture. Doing so, however, is not as simple as outsourcing the production of a new unit for either the OEM or the contract manufacturer. As will be explained in Chapter 7 (Production Planning in Remanufacturing), remanufacturing operations are quite different from the operations for producing new products, and this unfamiliarity by both parties may partially explain why more remanufacturing does not take place.

2.2.1 Direct Cost of Remanufacturing

So what other factors belong to the remanufacturing profitability assessment and what makes this analysis difficult? Besides the unfamiliarity problem, there are often significant costs associated with the logistics of remanufacturing. Remanufacturing involves the collection and transportation of the used units from the markets where they were sold to the location where remanufacturing processing takes place and then, transporting the remanufactured products to the markets where they will be sold. If we take the common case where an OEM's primary market is in North America and Europe but its contract manufacturers are primarily located in low-cost areas such as Asia, then the logistics cost of just shipping the core units across the ocean twice may be significantly higher than the new unit production case. Added to this is the cost of actually collecting the old units from the customers, who may be widely dispersed across a region and even unwilling to incur the hassle of facilitating the return of their used units without some kind of monetary incentive. The field of study that looks at how used products can be collected and where they should be processed to achieve the lowest cost is called reverse logistics, and is reviewed in Chapter 5 (Designing the Reverse Logistics Network). Because of the unfamiliarity in both remanufacturing operations and reverse logistics, simply quantifying a marginal "cost to remanufacture" may be a daunting task for many firms making it difficult to evaluate the remanufacturing business model.

The discussion above should make it apparent that it is very difficult to separate a firm's strategic decisions around remanufacturing, such as whether they should or should not remanufacture, with the more tactical decisions dealing with how the old units should be recovered and the remanufacturing operations should be run. Clearly, the answers to these tactical questions significantly influence the marginal cost of remanufacturing, which ultimately determines whether or not it is profitable to remanufacture. At the same time, a firm's strategy should also influence the tactical decisions associated with remanufacturing. For instance, Guide et al. (2006) describe how HP started thinking of its consumer-returns processing plants as profit centers rather than cost centers. By emphasizing faster turnaround times (extra capacity with lower utilizations) rather than minimizing cost (batch processing with longer lead times, higher utilizations), they were able to return remanufactured units to the market faster, before their market values had time to significantly decrease; an especially significant factor due to the short product life cycles inherent in the consumer electronics industry. For the remainder of this chapter, we will assume (a rather strong assumption for most companies) that we have a good idea of what the marginal cost of remanufacturing is. That is, for every unit that we remanufacture, what is the average cost of collecting the used product from the customer, sorting out and disposing/recycling of the seriously damaged units, transporting the remaining units to a processing location, testing and remanufacturing the units up to a "good-as-new" functional quality level, and transporting them to a location where they can be marketed to a customer? Even after knowing this important value, the evaluation of remanufacturing profitability is not as straightforward as it may first appear.

2.2.2 Opportunity Cost of Remanufacturing

So what may be missing from a simple per-unit price minus cost assessment of remanufacturing? Basically, this simple calculation does not include the opportunity costs of offering (or not offering) remanufactured products. One source of opportunity cost occurs when there are multiple potential uses of returned products. One alternative use for the old products is to harvest them for spare parts. Ferguson et al. (2008) argue that IBM, an OEM that actively remanufactures their used IT equipment, may sometimes make higher profits if they were to divert some of their returned cores to use for parts harvesting rather than remanufacturing. The reasoning is that even though their remanufactured products (such as laptop computers) often provide a higher margin than the spare parts that can be harvested, the remanufactured products also face more market uncertainty than the more stable demand for spare parts. Thus, they provide a model that explicitly makes this trade-off. The model uses the same basic principles that an airline uses when making the decision of how many seats on an aircraft to reserve for future potential higher-fare customers when there is ample current demand from lower-fare customers. In IBM's case, the current demand for spare parts represents the low-fare customers and the future, and more uncertain, demand for remanufactured products represents the high-fare customers. Even for firms that do not have any needs for spare parts, the option of recycling the returned units often provides another (possibly profitable) alternative versus remanufacturing and should be included in the decision-making process.

By far, the opportunity cost of remanufacturing that most firms seem to be concerned about is the fear that sales of the remanufactured product will reduce the demand for the firm's new product, commonly referred to as product cannibalization. Often, this claim is made without any substantial testing to back it up. Even some firms that are actively involved in remanufacturing their products often find their efforts restricted internally because of cannibalization fears. Restrictions put in place may include floors on the prices that can be charged for the remanufactured products (e.g., not be *x* percent of the new product price), limits on the markets where they can be sold (e.g., underdeveloped countries), the distribution channels the products can be sold through (e.g., outlet centers), the warranties that can be offered on them (e.g., half the length of a new product warranty), and the type of products that can be remanufactured. As an example of the latter type of restriction, Bosch Power Tools restricted remanufacturing to products where the firm had less than a 50 percent market share. Their thinking behind this rule of thumb is that for products with more than a 50 percent market share, the remanufactured product will cannibalize sales of the firms' new product but if the market share is less than 50 percent then it will cannibalize the sales of their competitors. The disadvantage of this strategy is that the firm does not benefit from selling remanufactured versions of its most popular products; something that is probably particularly objectionable to the manager in charge of remanufactured product sales.

Is this common fear of new product cannibalization justified? The debate on the extent of cannibalization by remanufactured products is still an ongoing one, by both industry experts and academics. One academic study seems to indicate that cannibalization may not be as much of a problem as some companies fear. Guide and Li (2007) listed the exact same versions, including the same warranties, of a power tool (consumer product) and an Internet router (business product) for sale on the eBay auction site but listed one as a new product and the other as remanufactured. They found that the bidders for the remanufactured power tool did not overlap with the bidders for the new power tool, even though the specifications for the tool were exactly the same. For the router, the bidding pools did overlap, but the people who bid on both the new and the remanufactured versions of the router tended to start bidding on the new router until it reached a certain price point and then switched to bidding for the remanufactured version. Thus, an argument could be made that these buyers would not have bought the new router because the final selling price went above what they were willing to pay for the product, so no cannibalization occurred. Performing similar experiments using Internet auction sites is one way for a firm to gage the degree of cannibalization a remanufactured product may cause. Of course, the argument that remanufactured products do cannibalize sales of new products has merit as well. The author of this chapter frequently purchases remanufactured laptops and electronic equipment in place of new versions of the same product, even though he has the means and willingness to purchase new products if the remanufactured version was not available.

Thus, the truth probably lies somewhere in the middle; remanufactured products do cannibalize the sales of new products but probably not to the degree that many firms fear.

2.2.3 Opportunity Cost of Not Remanufacturing

While most firms are aware of the opportunity cost associated with remanufacturing (especially the fear of cannibalization), there seems to be less attention paid to the cost of not remanufacturing. These less-familiar opportunity costs can often dominate the opportunity costs mentioned above, but as they are seldom considered, firms may make the (possibly) erroneous decision that remanufacturing is not profitable in their business. So what are the opportunity costs of not remanufacturing? First, there is the danger that ignoring the environmentally irresponsible product disposal practices of the firm's customers, a firm can find itself facing costly regulatory restrictions and government-mandated producer disposal fees in the future. This is already occurring in the electronics and automotive industries, with one regulation requiring that a certain percentage of each automobile be recyclable (European Union End-of-Life Vehicle Directive) while other regulations impose that electronic equipment producers fund the take-back and proper disposal of their products (WEEE). Research on how firms should (and do) respond to regulations such as these is reviewed in Chapter 3 (Environmental Legislation on Product Take-Back and Recovery). For our purposes, it is sufficient to acknowledge that operating an active and substantial remanufacturing program could reduce the risk of increased environmental legislation that mandates costly and possibly inefficient requirements on the OEM producer. Related to this potential benefit of being viewed as more environmentally friendly from a legislation standpoint, a firm may also achieve a benefit by obtaining access to a new market segment. Atasu et al. (2008) explore this possibility by modeling a "green" segment of customers who prefer a remanufactured product over a new product.

The possibility of costly environmental legislation (or the loss of the environmental market benefit) is not the only opportunity cost of not remanufacturing, however. Suppose an OEM determines that, in the absence of any opportunity cost, remanufacturing is profitable, but the decrease in profits of its new units, caused by the cannibalization of sales from the remanufactured units, exceeds the new profits available from producing and selling the remanufactured units. In this situation, the OEM may decide to ignore the (locally) profitable remanufacturing opportunity. By doing so, however, the OEM is leaving unclaimed older units (commonly referred to as cores) on the market that can be collected or purchased by a third-party firm. A third-party firm does not sell new units and thus, does not face the cannibalization opportunity cost of selling remanufactured units. Therefore, the third-party firm may find it profitable to remanufacture even though the OEM did not. This is exactly the case that has happened to many firms in the IT and printer industries. The market for refilled laser printer and inkjet cartridges provides a great example. Because of the high margins made by selling new printer cartridges, all of the major printer OEMs chose not to offer remanufactured (or refilled) cartridges for fear of cannibalizing this very profitable market. Of course, there are now thousands of third-party firms around the world that do offer refilled cartridges, much to the dismay of the major OEMs (Lyra Industries Reports 2008). In response, the major printer OEMs have waged an ongoing fight (mostly unsuccessfully) against the third-party cartridge refillers using lawsuits, new technology, frequent design changes, and threats of invalidating product warranties when refilled cartridges are used. When a situation like this occurs, the OEM is often worse off than if they had chosen to remanufacture themselves; they still incur the cannibalization of their new product sales but a third-party firm is reaping the profits from selling the remanufactured units rather than the OEM. The inclusion of this opportunity cost in the OEM's decision of whether or not to remanufacture is explored in Ferguson and Toktay (2006).

The loss of profits from the remanufacturing business is not the only concern of OEMs when third-party firms sell remanufactured versions of the OEMs' products. In comparison to the markets for new products, which typically consist of a small set of large OEMs, the third-party remanufacturer market is very fragmented and often made up of many small- to medium-size firms. For example, in the IT networking industry, there are over 300 firms whose primary business is selling refurbished networking equipment (www.uneda.com). This is mainly because the barriers to entry are rather small for most type of products; they do not require significant capital investments to set up a remanufacturing operation. Another reason the market for third-party remanufacturing firms tends to be fragmented is that it is difficult for a firm to build a brand name when the product the firm sells still carries the brand name of the OEM. Thus, the brand image of the name appearing on the remanufactured product is often valued higher by the OEM that originally produced the product than a small third-party firm that remanufactured the product. As a consequence, the quality standards required by the third-party firm may not be as high as the OEM would like them to be. Of course, low-quality remanufactured products hurt the entire remanufacturing industry as well as the brand image of the remanufactured product's OEM. To try to minimize this hit to the industry's reputations, reputable third-party firms and OEMs sometimes form alliances and create certification programs that ensure remanufactured products meet some minimum quality standard. For example, IBM offers a low-cost certification of remanufactured IBM equipment to third-party firms, where an IBM engineer will inspect the remanufactured product and give it a seal of approval. Programs such as these often create an uneasy dilemma for the OEM, however, because, on the one hand, the OEM wants its customers to perceive the remanufactured products as low-quality substitutes to the OEM's new products, but on the other hand, many customers will attribute the poor quality of a remanufactured product to the OEM's name on the product, even when a third-party firm performed the remanufacturing. Thus, the added difficulty of maintaining a high-quality brand image is another opportunity cost faced by an OEM who chooses not to remanufacture (thus making it easier for third-party firms to enter the market).

Now suppose the OEM is in an industry where either the capital investment required to set up a remanufacturing operation is prohibitively high or the OEM has some means to control the profitability of any third-party remanufacturer firms. The former is the case for very capital-intensive products such as jet aircraft engines and the latter is the case for IT OEMs that sell products such as servers and routers that require specialized software. For the case of the IT OEMs, they have created a powerful mechanism that allows them to exert control over the secondary market of their products. The mechanism is a software relicensing fee that is required from any new owner (other than the original purchaser) to legally use the software installed on the server or router. Thus, by setting a large-enough relicensing fee, the OEM can make it unprofitable for third-party firms to sell a remanufactured version of its product as the customer of the remanufactured product will also have to pay the relicensing fee to the OEM to be able to use the product.

So is there still an opportunity cost of not remanufacturing for OEMs in these types of industries? Some clue may be given by observing the different practices of OEMs in the same industry. For example, in the IT server industry, Sun Microsystems has historically charged a relicensing fee of up to 70 percent of the new product's selling price; essentially eliminating any secondary market for their product. In contrast, IBM, who sells servers that are close competitors to those of Sun, only charges a relicensing fee of around 1 percent of the new product selling price. Something must be driving these radically different secondary market strategies. As IBM and Sun most likely face similar environmental legislation pressures and cannibalization costs, there must be an additional opportunity cost we are missing. This new opportunity cost takes the form of a "resale value effect," where a forward-looking customer will take into account the price that a product can be sold at on the secondary market when making his or her initial buying decision. A good analogy for this occurs in the automobile market. Suppose you are choosing between two cars (Brands A and B) with similar performance and both priced at $25,000 but you know that you can resell Brand A in two years for $10,000 but there is no secondary market for Brand B. Clearly, you would choose Brand A, even if you planned on keeping the car for all of its useful life: the Brand A car gives you more options for the same performance and price. If you knew you only needed a car for two years, you may even be willing to pay up to $10,000 more for Brand A over Brand B. This opportunity cost is explored in Oraiopoulus et al. (2008), who show that when customers are forward looking, it is never optimal for the firm to set a relicensing fee so high that they completely shut down the secondary market. What may explain the difference between Sun and IBM's strategies then is that, historically, IT customers have not thought much about resale values as secondary markets for IT equipment were not well developed. This started to change, however, during the dot-com bubble of the late 1990s when shortages for new IT equipment created a demand for used IT equipment that was quickly filled by many small refurbishers. Today, there is a substantial secondary market infrastructure in place and prices for used IT equipment are easily available. Thus, it is much more common today for the purchaser of new IT equipment to look up the resale value of a product before purchasing. Sun seems to have belatedly recognized this trend (after a prolonged loss of market share) and has recently significantly lowered their relicensing fees.

2.3 Conclusion

Strategic issues in closed-loop supply chains involve high-level decisions such as whether or not OEMs should participate in, support, or even try to deter the secondary market of their products. At first glance, the decision seems like it should reduce to a simple profitability analysis: if it is profitable to collect their used products, possibly remanufacture them, and sell them for a profit, then the firm should do so; if not, then the firm should not participate in the secondary market (refurbished or remanufactured product market). For firms that have not previously been involved in used-product collection or remanufacturing, even this direct calculation is challenging because of the difficulty of estimating the costs of collection and remanufacturing. Thus, the strategic question of whether or not to be actively involved in the secondary market is intricately linked to the more tactical questions of how to set up a collection system and plan a remanufacturing production process. There already exists a substantial amount of work that addresses these tactical issues but the more strategic question of should a firm remanufacture has only recently received attention in the academic research community. In this chapter, we summarize some of these recent works and argue that there are several opportunity costs associated with a firm's decision to participate (or not) in the secondary product market. These costs are rarely quantified and often even not considered when firms make their strategic decisions. Thus, our goal is to increase the awareness of these opportunity costs so that firms can make more informed decisions regarding their involvement in the secondary market for their products. More specifically, we discuss some of the opportunity costs associated with a firm's decision of whether or not to remanufacture their used products and resell them in the secondary product market.

The opportunity costs associated with remanufacturing include the loss of the used products for other uses such as recycling or harvesting for spare parts and the potential cannibalization of the sales for the firm's new products. The former is often an issue for manufacturers of complex products such as computers, engines, construction equipment, or industrial machinery. Because firms in these types of industries have developed intricate spare parts supply chains, they may not have systems in place to allow them to recognize the true value in meeting their spare parts needs through the harvesting of returned products. Thus, a firm may give priority to remanufacturing all returned units over a specified quality level rather than solving for the right balance between remanufacturing and parts harvesting. The latter opportunity cost is typically well recognized by firms, but is rarely empirically tested and quantified. Indeed, just the fear of any cannibalization of the firm's new product sales from the selling of (lower priced) remanufactured products is enough to deter many OEMs from remanufacturing their returned products.

On the other side, the opportunity costs associated with not remanufacturing include the cost of future potentially expensive legislation, the cost of leaving a profitable market open to a third-party firm that may remanufacture and resell your used products, and the reduction in the customers' value for a new product due to the resale value effect. The threat of future expensive environmental legislation is increasing in importance and awareness, especially in the electronics industry where several countries and some states within the United States have already passed laws that hold the OEMs responsible for the proper end-of-life disposal of their products. Unfortunately, the most common response by firms to the threat of this type of legislation is to spend money lobbying for more favorable legislation than to invest in an environmentally sound secondary market strategy that minimizes the need for such legislation. The opportunity cost of allowing the entry of third-party remanufacturers is often even more detrimental to an OEM's long-term profits. Common responses by OEMs to this new category of competition is to dismiss the quality of the remanufactured products as being substandard, or not supporting the remanufactured products through voided warranties or costly relicensing fees. Such tactics, however, lead to the third opportunity cost—reducing the customers' valuation of a new product due to the absence of a healthy secondary market where the customers can sell their used products. Thus, firms that actively try to deter the secondary market for their products may hurt their overall long-term profits by doing so.

References

Atasu, A., M. Sarvary, and L. N. Van Wassenhove. 2008. Remanufacturing as a marketing strategy. *Management Science* 54, 1731–1746.

Berko-Boateng, V. J., J. Azar, E. De Jong, and G. A. Yander. 1993. Asset recycle management—A total approach to product design for the environment. In *International Symposium on Electronics and the Environment*, Arlington, VA, IEEE.

Ferguson, M. and B. Toktay. 2006. The effect of competition on recovery strategies. *Production and Operations Management* 15, 351–368.

Ferguson, M., Fleischmann, M., and G. Souza. 2008. A Capacity-Based Revenue Management Approach to Disposition Decisions in Reverse Supply Chains. *Working Paper*. College of Business, Georgia Institute of Technology, Atlanta, GA.

Guide Jr., V. D. and K. Li, 2007. The Potential for Cannibalization of New Product Sales by Remanufactured Products. *Working Paper*. Smeal College of Business, The Pennsylvania State University, Philadelphia, PA.

Guide Jr., V.D., G. Souza, L. N. Van Wassenhove, and J.D. Blackburn. 2006. Time value of commercial product returns. *Management Science* 52, 1200–1214.

Gutowski, T. G., C. F. Murphy, D. T. Allen, D. J. Bauer, B. Bras, T. S. Piwonka, P. S. Sheng, J. W. Sutherland, D. L. Thurston, and E. E. Wolff. 2001. *Environmentally Benign Manufacturing*. Baltimore, MD: World Technology (WTEC) Division, International Technology Research Institute.

Hauser, W. and R. Lund. 2008. *Remanufacturing: Operating Practices and Strategies*. Boston, MA: Boston University.

Lyra Industry Reports. 2008. The State of the Aftermarket Printer Supplies Industry: Overview and Analysis. Available at http://lyra.ecnext.com/coms2/summary_0290-901-ITM.

Oraiopoulus, N., M. Ferguson, and L. B. Toktay. 2008. Relicensing Fees as a Secondary Market Strategy. *Working Paper*. College of Management, Georgia Institute of Technology, Atlanta, GA.

Toktay, B., L. Wein, and S. Zenios. 2000. Inventory management of remanufacturable products. *Management Science* 46, 1412–1426.

U.S. EPA. 2007. Office of Solid Waste: Basic Facts. Available at www.epa.gov/garbage/facts.htm.

U.S. EPA. 2008. Office of Solid Waste: Reduce Reuse and Recycle. Available at www.epa.gov/epaoswer/non-hw/muncpl/redulce.htm.

Valenta, R. 2004. Product Recovery at Robert Bosch Tools, North America. In *Presentation at the 2004 Closed-Loop Supply Chains Workshop*, INSEAD, Fontainebleau, France.

Chapter 3

Environmental Legislation on Product Take-Back and Recovery

Atalay Atasu and Luk N. Van Wassenhove

Contents

3.1 Introduction

This chapter aims to provide a business perspective on how environmental legislation affects manufacturing systems and operations. We focus on the extended producer responsibility (EPR) approach, which holds producers/manufacturers physically and financially responsible for the environmental impact of their products after the end of life. Our examples are generally based on the electronics industry, as the diffusion of environmental legislation is the fastest for this industry in today's economy.

Over the past ten years, legislators in different parts of the world have adopted the principles of EPR and implemented legislation that enforces manufacturer responsibility for environmentally responsible treatment of products that reach the end of their useful lives. The waste electrical and electronic equipment (WEEE) and end-of-life vehicle (ELV) directives in Europe, and The Specified Household Appliance Recycling (SHAR) Law in Japan have been some early examples of such legislations. While the European Union and the Japanese government pioneered, a number of states from the United States followed. Starting from 2004, 12 states (CT, ME, MD, MN, NJ, NC, OK, OR, TX, VA, WA, WV) passed e-waste bills mandating manufacturer responsibility for end-of-life products. Some states already started collection and recycling programs, while the majority of the programs are expected to start operating in 2009. A number of other states are known to be considering EPR legislation.

The existence and the diffusion of such legislation around the world raises the question as to what the goal of EPR is. From the legislator's perspective, the ultimate goal should be the reduction of the environmental impact by proper recycling and the disposal of e-waste while keeping the social-economic impact at a marginal level. In other words, EPR should maximize social welfare (including the environmental impact). The goal of manufacturers, on the other hand, is usually to comply with the law at the minimum possible cost. Consequently, certain conflicts, such as environmental benefits versus economic impact (increased costs), are inherent in the nature of EPR. Our purpose in this chapter is to lay down the basics and provide a better understanding of efficiency issues in such legislation from the business perspective.

We first review the environmental economics literature that investigates the impact of EPR on the society and the economy as well as its impact on the environment in an ideal world. We note that the focus of this literature is a social one and does not necessarily provide a business perspective. Nevertheless, it is important to understand the policymaker's perspective. Thus, we first discuss the policy models developed by this literature and find out what type of legislation would be the most efficient in an ideal world. After that, we look at the existing legislative models in practice and explain the basics and different approaches in EPR legislation implemented in different parts of the world. Finally, the last layer of our discussion considers recent business operations management (OM) articles that deal with the implications of EPR legislation on businesses and manufacturing economics. We conclude by identifying important factors that businesses have to take into account when facing EPR legislation.

3.2 What Do the Economists Say?

The environmental economists investigating EPR focus on how the socially optimum amount of waste generation and disposal can be ensured in stylized models of the economy (see Palmer, Walls, and Sigman 1997, Palmer and Walls 1997, 1999, Fullerton and Wu 1998, Calcott and Walls 2000, 2002, Walls and Palmer 2000, Walls 2003, 2006). Once again, the focus of this literature is a social one, problems are approached from the policymaker's perspective, and the goal is to attain the best social outcome.

One of the earliest economic models was proposed by Palmer, Walls, and Sigman (1997), who compare the social costs of three different policies (deposit/refund system, recycling subsidies, and advanced disposal fees) in reducing municipal solid waste and conclude that deposit/refund is the least costly policy. Similarly, Palmer and Walls (1997) discuss the efficiency of deposit/refund systems and recycling content standards in generating the socially optimum amount of disposal. Both use partial equilibrium models with competitive markets and do not take into account product recyclability in their analysis.

Fullerton and Wu (1998) and Walls and Palmer (2000) formulate models that take into account all environmental externalities throughout the whole life cycle of a product. In this setting, they discuss the efficiency of various upstream and downstream policies (e.g., disposal fees, subsidies on recyclable design, command and control regulatory standards, deposits and refunds) in ensuring the socially optimum level of product recyclability. They conclude that depending on the objectives, market failures, and the ease of implementation, different policies can be useful in obtaining the social optimum. Calcott and Walls (2000, 2002) also investigate the success of deposit/refund systems and disposal fees in encouraging design for environment (DfE) and product recyclability. They conclude that downstream policies (e.g., disposal fees, taxes imposed on products) are not useful or practical in encouraging product recyclability, especially considering the lack of fully functional recycling markets. They show that deposit/refund-type policies can be more effective in obtaining the constrained social optimum of recyclability.

Unlike Fullerton and Wu (1998), Calcott and Walls (2002) explicitly consider a recycling market where instead of simply assuming that these markets either function or not, they argue that there may be some transaction costs that obstruct the functionality of the markets and analyze the effects of transaction costs on the efficiency of the environmental policies. Palmer and Walls (1999) and Walls (2006) use case studies to discuss the pros and cons of different environmental policies. Palmer and Walls (1999) examine three specific policies (upstream combined product tax and recycling subsidy (UCTS), manufacturer take-back requirements, and unit-based pricing) and conclude that UCTS, which is a special type of a deposit/refund system, is more cost effective, especially in terms of transaction costs. Walls (2006) provides a more extensive overview and comparison of various policies under the EPR umbrella and presents insights from real-life applications of these policies.

The common features of all these studies are the focus on the social optimum and to what extent it can be attained by different policies and the consideration of product recyclability and DfE decisions. Below, we provide a summary of policy tools put forward by this literature and identify their strengths and drawbacks. Walls (2006) is extremely useful in this sense as the author compares the effects of various policy instruments on possible objectives of EPR, namely, advance recovery fees, recycling subsidies, unit-based pricing, take-back mandates, and recycling rate targets.

An advance recycling fee (ARF) is a fee collected from consumers or producers for recycling of the products they purchase or sell. Consumers pay this at the time of purchase and the producers are charged on product sales. Generally, in an advance recycling fee (ARF) system (see the California and Taiwan examples in the next section), producers or consumers are charged per product or unit weight sold. Walls (2006) states that with ARF, production and consumption are expected to decrease and thus, less virgin material would be used. If ARF is charged per unit weight of the product, then product design can be slightly affected as producers try to reduce the size and the weight of their products.

In a recycling subsidy system, the recycling party is paid a per item subsidy. In such a model, product design is indirectly influenced by subsidies. Production and consumption are expected to increase and greater output offsets the reduced usage of virgin materials. Recycling is improved and all these effects are larger when the subsidy is granted based on product weight rather than per unit weight. This instrument needs funding from the social planner side, which makes it harder to implement.

In a deposit/refund system, a tax on production or consumption is associated with a subsidy proportional to product recyclability. A recycling subsidy, when combined with ARF, is an example of such a system. This would directly improve recycling and reduce virgin material usage and product consumption. It also helps in reducing the product weight and improving DfE. Further, the financing of subsidies can be handled through the advance recycling fees collected.

A recycling target is a standard recycling rate set by the policymaker and can be defined as the proportion of product sold that needs to be recycled. In tradable recycling credits scheme, if a producer is unable to achieve the target recycling rate, he can buy equivalent credits from other firms at a price. Similarly, a producer can sell unneeded recycling credits to firms who need them. Such regulation gives incentives for producers to reduce product size and weight. It may also reduce output and virgin material usage. Recycling is increased as a result, but it needs a producer responsibility organization to take care of take-back operations, which is a cost addition. When this scheme is combined with a tradable credit scheme, it has a more direct effect on product design, but transactions could be costly.

A unit-based fee policy charges the end user for the cost of recycling. Such a model, that is, pay-as-you-throw policy, reduces output quantity and the virgin material usage. It also indirectly improves recyclability. The main disadvantage of this instrument is that it can lead to illegal dumping.

Some of these policy instruments provide optimum amounts of waste disposal and recycling, but need extensive effort from the government in monitoring and documenting the critical environmental characteristics of products like their recyclability. So, the question to be raised is how practical these tools are? In the next section, we look at the practical situation for the electronics industry in a variety of geographical locations.

3.3 What Is Happening in Practice?

Having discussed the suggestions by the economists, we now look at what is happening in practice, focusing on the electronics industry. In all of our practical examples, we observe that three categories of policy tools are employed, namely, recycling targets, advance recycling fees, and unit-based fees. This is interesting given that the previous discussion from the environmental economics literature suggests that all three policies have drawbacks and fail to attain the social optimum. The question then is why these policies have been chosen. Perhaps a practical explanation is the difficulty of implementing more complicated policies. It may be costly to operate and monitor policies with multiple levers such as the deposit/refund model. Similarly, industry dynamics and lobbying may be very influential on how the policy instruments are chosen. The process underlying the policy decisions should of course be an important concern to businesses; however, it is not a core question to this chapter. Our goal is to focus on explaining how existing systems operate, what differences exist between those, and how they can be improved. We proceed with a detailed description of some models, which we believe cover a broad range of differences between current legislations.

3.3.1 The WEEE Directive in the EU

Our first example has probably the largest scope compared to the legislation in other parts of the world. The WEEE Directive (Directive 2003/108/EC) enforces producer responsibility for end-of-life electrical and electronic waste in Europe.

Producers are physically and financially responsible for meeting certain recycling or recovery targets, while the member states must guarantee that 4 kg of such waste is collected per capita per year, at no cost to the end users.

An important deficiency of the WEEE Directive, from the industry perspective, is the collective nature of cost allocation between manufacturers. The WEEE Directive clearly states that producers should be allowed to have access to their own waste and only be responsible for their waste. However, in a significant number of countries, manufacturers are required to join collective systems where the cost allocation is based on market shares (Belgium [Walloon], Denmark, Estonia, Finland, France, Greece, Latvia, Portugal, Slovenia, Spain, and the United Kingdom). This is widely criticized by a number of manufacturing organizations because market shares are not necessarily a good representation of waste shares, and there are significant differences in the recovery costs between manufacturers. For example, having a cell phone manufacturer and a computer monitor manufacturer share recovery costs based on market share is not a fair system. As recovered cell phones can generate additional profit and monitors are costly to recycle, it is not in the cell phone manufacturer's interest to share the monitor manufacturer's recycling costs. This explains why a number of manufacturing organizations lobby against the collective systems and demand individual producer responsibility. The opposing point of view argues that collective systems are beneficial due to economies of scale, that is, the average recovery cost would be lower when larger volumes are recovered.

Atasu and Boyaci (2009) argue that another significant difference between collective and individual systems must be about cost efficiency. Collective systems are expected to result in higher costs on the average, even more in a monopolistic system. Another important difference, according to the authors, concerns design incentives. Atasu and Subramanian (2009) show that individual systems are likely to generate superior incentives for recyclable product design.

Our next examples are from the United States. Although there are currently 13 states that have enacted product take-back legislation for electronics, we focus our discussion on the examples of Maine, Washington, and California for the sake of brevity.

3.3.2 United States: Maine and Washington

The producer responsibility directives in Maine and Washington cover household consumer products such as computers, televisions, and DVD players. Maine's directive has been in effect since January 1, 2006. Washington's directive came into effect on January 1, 2009. The directives generally resemble the WEEE Directive, but an important difference is that the Maine and Washington directives use the "return share" model, where manufacturers pay for the recycling costs associated with their share of products in the waste stream. Manufacturers consider the return share model to be a step closer to the individual responsibility concept as compared to the market share model. The return share model is also being contemplated for the planned product recovery programs in the states of Connecticut and Oregon.

In the Maine directive, the collection task is assigned to the municipalities, who then pass the waste to one of the seven previously assigned consolidators. The manufacturers have to arrange the collection and recycling systems. Two options are allowed: (1) they can collect a proportion of waste (based on their return share) and recycle it or (2) they can have a consolidator recycle their share. The manufacturers' return share is calculated by statistical sampling from the waste stream.

The Washington directive is similar to the Maine directive. It requires manufacturers to participate in an approved recycling plan as defined by the state. Manufacturers may join a collective system, which is called the standard plan. They can also act individually, as long as their plan conforms to the standards in the legislation. Finally, they can join a collaborative system with other manufacturers. Cost allocation for collective or collaborative plans should be based on return shares of the manufacturers. The Department of Ecology (DE) has implemented a system (Brand Data Management System developed by the National Center for Electronics Recycling [NCER]) to calculate the return shares of each manufacturer.

3.3.3 United States: California

California is the first U.S. state to establish an advance recycling fee program. The Californian legislation charges consumers an advance recycling fee at the moment of purchase of a product that contains a screen. The fee varies between $6 and $10, depending on the size of the product. The fee applies to all sales of displays with a diagonal screen size of at least 4 in. The fee is $6 for screens between 4 and 15 in., $8 for screens between 15 and 35 in., and $10 for screens larger than 35 in. The fee applies to all transactions in which the California sales tax applies, including leases, and to Internet and catalog sales to purchasers who take possession in California. Failure to collect the fee is punishable by a fine of up to $5000 per sale. The local governments use part of the advance recycling fee to subsidize authorized collectors and recyclers, while 3 percent of the fee is kept by the retailers. Under the act, manufacturers must provide consumers information regarding recycling opportunities and, since July 1, 2005, must report to the California Integrated Waste Management Board on the number of covered devices sold and the amount of hazardous materials they contain.

3.3.4 Taiwan

The Taiwanese Scrap Computer Management (SCM) Foundation, which was established on June 1, 1998, supervises the operation of the computer recycling program. It collects a processing fee from the manufacturers and importers of computers. This fee is collected per recycled item. Currently, the scrap computer processing fees for the designated items are as follows: PC main printed circuit boards NT$75/unit, PC hard disks NT$75/unit, PC power suppliers NT$12.5/unit, PC frame shells NT$12.5/unit, PC monitors NT$125/unit, and notebook computers NT$200/unit. The legislation states that these processing fees will be recalculated according to actual costs and, if necessary, reset at the end of each year. However, the fees have remained unchanged since 1998.

The SCM Foundation also offers reward money for consumers who bring their unwanted computers to designated collection points to increase consumer participation. These collection points mainly consist of computer retailers who are in a good position to receive scrap computers from consumers. The collection points can provide consumers reward money on the spot and receive rewards from doing so. Collection points receive NT$50 for every notebook computer, NT$60 for every PC mainframe, and NT$70 for every PC monitor. Currently, recyclers are subsidized from a budget funded by the disposal fees. The disposal fees are fixed at the numbers given above, while the associated subsidies are determined on the basis of breakeven between revenues and costs along with recycling operations. The efficiency of this approach is still unknown, however, given the fact that sales/disposal ratios vary throughout the product life cycle.

3.3.5 Japan

Our interactions with practicing managers in the electronics industry suggest that one of the most favored take-back legislations has been enacted in Japan. The Japanese directive, which started in April 2001, sets treatment standards via a waste management law. The directive's scope is limited to TV sets, cooling devices, washing machines, and air conditioners. It assures that end users pay for the end-of-life management of products through a return share system. End users are charged an end-of-life management fee by the manufacturer upon disposal that is collected by the retailers and used for the management of a common recycling center. The Japanese system is capable of distinguishing brands and properties of products. Each producer has control over the fate of his products, that is, recycling, repairing, etc., and both producers and end users have the possibility of tracking where the products are treated through a so-called manifest system. The manifest system also enables the recycler to identify the producer of the product through the recycling flows. Using a so-called recycling bill, the system identifies applicable collection points and recycling plants according to the brand and the category of the product. It allows for statistical data collection and ensures the traceability of individual waste products and responds to customer inquiries.

The advantage of such a system is that it allows the manufacturer to get feedback about the end-of-life issues related to the product. The recycling plants provide the manufacturer with product design–related feedback from the recycling of their own product. Feedback reports from the recyclers cover proposals for design improvements on issues such as material composition, ease of disassembly, and labeling. The striking feature is that the Japanese system creates incentives for greener designs.

Incentives to improve the efficiency of recycling operations create positive feedback on greener designs, sometimes even beyond the legal requirements.

3.3.6 Sweden

Finally, we go back to Europe to provide an example of a national legislation (independent from the legislation of the European Commission) that specifically targets green design improvements. Sweden has established a unique financial system that guarantees the recycling of cars at their end of life. As part of this system, automobile manufacturers pay negotiated insurance premiums to a private insurance company to cover future recycling costs when automobiles return for recycling. Premiums are based on estimates of future recycling costs. The principal benefits of this system have been identified as (1) the mitigation of uncertainty in future recycling costs and (2) incentives for environmentally better designs because superior recyclability results in lower insurance premiums. Similar premiums are offered by Swedish insurance companies for electronic waste.

3.3.7 Discussion

The examples cited above show that there are additional complexities embedded in EPR legislation. Although similar tools, such as recycling targets or unit-based fees, may be used for policymaking, the implementations in different countries differ significantly. As one would expect, implementation-related differences may lead to different outcomes, cause disturbance in competition, and create fairness concerns. Our experiences with practicing managers suggest that this is the case. While some implementations are favored by a group of manufacturers, others prefer alternatives. This basically leads to the suggestion that to anticipate the impact of such legislation at social, business, or company levels, one has to clearly understand the impact of the exact implementation structure. This requires systematic analysis of such systems. One way to do this is to factor out some important effects as follows:

1. What policy tool is chosen? Recycling rate, advance recycling fee, or unit-based fee?
2. Recovery management: Is there a single compliance scheme, or is there competition in the recycling market?
3. Physical responsibility: Are manufacturers collectively or individually responsible?
4. Financial responsibility: Who has the financial obligation: the end user, the purchaser, or the producer?
5. Cost sharing: If a collective producer responsibility system is employed, how is the cost allocation made between producers? Is it based on market share or return share? Is there recycling cost differentiation between producers?
6. Design incentives: Does the EPR legislation provide incentives for recyclable product?

Although we believe that this list is extensive and covers most of the practical issues to date, it is hard to come up with a best-case scenario for all types of producers and industry environments. Thus, a more pragmatic approach is needed to understand how such factors drive the efficiency of EPR legislation and how different business environments are affected. In what follows, we discuss a few recent academic papers that provide technical tools to shed light on the business implications of different legislative models.

3.4 What Is the Operations Management Perspective?

A few OM papers have appeared recently, mainly looking at production economics and competition under EPR legislation from the business perspective. We provide a detailed summary of each below. Our purpose is to understand how take-back legislation impacts production economics, competition, product design, and basics of supply chain management. It is very important to note that the factors mentioned in the previous section, (e.g., policy tools, responsibility assignments, and modes of cost sharing) play significant roles in the impact of legislation.

3.4.1 Production Economics

The first article in our discussion (Atasu et al. 2009a) provides a bird's-eye view on the general drivers of economic efficiency of take-back legislation. The authors use a generic model of the economy to analyze the environmental and economic impacts of environmental legislation similar to the WEEE Directive.

The authors consider a competitive market place, and show that the social planner should set target collection levels according to the intensity of competition in a market. They also show that reducing the environmental impact is always a benefit to a monopolist, provided that the legislation sets targets according to environmental impact. With those observations, the authors come up with the following policy suggestions: (1) Weight-based take-back legislation may not be using an efficient measure of cost to the environment. (2) Legislation should set collection and recovery targets based on the environmental characteristics of the products. However, an interesting observation concerns manufacturer reactions to legislation under competition: In a WEEE-like legislation, a manufacturer with lower environmental impact is punished for other manufacturers' environmental hazards. A manufacturer, by reducing the take-back related costs, can lower other manufacturers' profit and increase his profit. Hence the manufacturers may tend to decrease treatment costs instead of increasing the environmental quality of the product. This finding signals the importance of individual producer responsibility for the sake of fairness and green designs. The problem of fairness can be resolved by making every single manufacturer responsible for their own products, that is, by IPR. Furthermore, individual responsibility models are likely to create better design incentives.

3.4.2 Policy Choices

The next article in our discussion (Atasu et al. 2009b) investigates the impact of policy choices on production economics. The authors extend the model given in Atasu et al. (2009a) to account for the impact of different policy tools. They argue that structural differences in the existing legislation would impact the welfare of different stakeholders differently. They focus on existing EPR models and observe that they can be classified in two categories based on the policy tool used: (1) a tax model and (2) a recovery target (rate) model. In the tax model, the social planner charges manufacturers (or consumers) a unit tax per item and undertakes the collection and recovery tasks. Thus, in this model, manufacturers or consumers are only financially responsible for end-of-life products. In the second model (denoted as the rate model from now on), the social planner sets certain collection or recovery targets, and the manufacturers are both physically and financially responsible for end-of-life products.

The main knowledge from this article is that a naive social welfare maximizing solution is a tax model that supports the typical argument made by most manufacturers in Europe. They believe that the current WEEE model (which essentially is a rate model according to Atasu et al. (2009b)) is designed to shift the burden of operating take-back systems from the government to the manufacturers. But the economic analyses of the two models show that this is not always true. Manufacturers can indeed benefit from the rate model even when the costs of operating the two systems are effectively the same. Given that potentially the manufacturers can further reduce the costs of operating their own systems as compared to the costs to be incurred under a government run system, the rate model can be even more beneficial for the manufacturers.

3.4.3 Cost Sharing within a Supply Chain

The third article we consider (Jacobs and Subramanian 2009) follows a similar model to Atasu et al. (2009a,b). Jacobs and Subramanian argue that EPR programs typically hold the *producer*—a single actor defined by the regulator—responsible for the environmental impacts of end-of-life products. This is despite emphasis on the need to involve all actors in the supply chain to best achieve the aims of EPR. Thus, they explore the impact of sharing EPR program costs between tiers in the supply chain. The authors demonstrate that social welfare is significantly affected by the interaction between the program cost–sharing level, the recovery rate, and the nature and the magnitude of the externality functions. Thus, the social welfare outcomes from sharing EPR program costs are intricate and care should be taken in designing them to ensure a balance between economic and environmental performance.

3.4.4 Supply Chain Coordination

The next article in our discussion (Subramanian et al. 2009) studies the influence of EPR policy parameters on product design and coordination incentives in a *durable product design supply chain*. Their focus is on studying the impact of supply chain coordination on design choices and profit. The authors show that the design choices of an integrated supply chain are environmentally superior to those of a decentralized supply chain. Thus, supply chain coordination can help improve the environmental quality of products. Furthermore, the authors investigate the impact of legislative parameters on the efficiency of the supply chain. They show that while disposal costs are usually aimed at reducing a product's end-of-life environmental impact, they can also help improve product designs so that the product's during-use environmental impact is reduced.

3.4.5 New Product Introductions

Following Subramanian et al. (2009), the next issue we would like to consider in this section is how take-back legislation influences product designs. One way to look at this problem is to understand whether the frequency of new production would change under take-back legislation. Plambeck and Wang (2009) argue that rapid or frequent new product introduction is harmful for the environment as it increases the amount of waste, as well as resource extraction, and they question the effect of take-back legislation on the frequency of new product introduction. The authors show that product take-back legislation would extend the useful life of the product and reduce the volume of e-waste by reducing the frequency of new product introduction, and this effectively increases manufacturer profits. Furthermore, such regulation can be more beneficial with more intense competition because manufacturers under competition are rushed by competitive pressures and would benefit from being slowed down by take-back legislation.

3.4.6 Design for Recycling

While Plambeck and Wang (2009) focus on new product introduction frequency, Atasu and Subramanian (2009) deal with the impact of take-back legislation on the environmentally friendly design choices, for example recyclability, of manufacturers. The authors consider a stylized market that consists of a differentiated duopoly, consisting of a high-end and a low-end manufacturer. With this model, the authors analyze how design incentives are created under collective or individual producer responsibility. The authors find that under a collective system, the equilibrium recyclability of high (low)-end manufacturers increases (decreases) in the proportion of high-end sales. Furthermore, the high-end manufacturers are less likely to choose a greater recyclability than the low-end manufacturers. On the other hand, when an individual producer responsibility model is employed, both high-end and low-end manufacturers design more recyclable products under individual responsibility than under collective producer responsibility. In other words, individual producer responsibility models are superior to collective models in terms of design implications. Furthermore, if synergies can be created, even better product designs can be obtained if the manufacturers collaborate under individual producer responsibility systems.

3.4.7 Recycling Markets

Finally, Toyasaki et al. (2008) investigate the impact of scale economies and recycling market competition on the efficiency of product take-back legislation. They observe that there exist two types of recycling markets in countries where product take-back legislation is enacted: monopolistic and competitive. Practitioners, for example, manufacturers or legislators, argue that monopolistic systems benefit more from economies of scale while competitive systems have the potential to reduce recovery costs due to recycling market competition. While the conflict between the two types of markets is clear, it is not known under which conditions one of the two market models dominates in terms of economic efficiency.

Comparing the recycling fees and profits under the collective and monopolistic take-back systems, the authors show that the average recycling fee in a monopolistic system is always higher than that in a competitive system. This means that the average manufacturer and consumer would be economically better off with competitive systems. In addition to this, the authors show that the monopolistic systems are more harmful for the low market share manufacturers in a differentiated competition model. This is because monopolistic systems impose a fixed recycling fee that is the same for all manufacturers. Thus, it is likely that low-end manufacturers obtain higher benefits from competitive systems.

3.5 Discussion and Conclusions

In this section, we provide an overview of what we learned from academic articles and practical implementations of product take-back legislation. First of all, we observe that socially optimal policy tools may not be preferred in practice because of implementation difficulties or focused lobbying of manufacturing organizations. Thus, businesses are likely to have the possibility to influence the implementation of take-back legislation that benefits them. However, because the objectives of different producers/businesses are different, it is hard to understand what types of policies favor whom. The preceding discussion is useful in developing an understanding of these issues.

First, we discuss the policymaker perspective on the efficiency of existing legislation. Atasu et al. (2009a,b) show that policymakers can improve social welfare even under suboptimal policies by optimally setting the implementation-related parameters. The policymaker can, for instance, use a collective producer responsibility system under certain conditions to improve the R&D choices and the profitability of the manufacturers (see Plambeck and Wang 2009), while an individual producer responsibility system can be used to improve the recyclability choices (see Atasu and Subramanian 2009). Similarly, the recycling market structure is something the social planner can influence. By choosing a competitive recycling market, social welfare and recycling levels can be improved (see Toyasaki et al. 2008). In conclusion, the social planners can anticipate the business reactions to structural differences in existing legislation and set up legislation in a more effective manner, even under suboptimal policy tools that are preferred due to their ease of implementation.

Next, we consider the manufacturer's perspective. Perhaps the most important concern raised by take-back legislation from the manufacturer's perspective is how their competition is affected as the assurance of fairness in legislation seems to be their most important concern. According to Atasu et al. (2009a), manufacturers should be aware of the target-setting mechanism and the collective nature of product take-back legislation. Although collective systems, where recovery targets/fees are based on the weight of the product, seem to be cost efficient, they are not necessarily fair. It can be argued that fairness concerns should outweigh the cost-efficiency concerns, and individual responsibility systems should be used along with environmental impact–based recovery targets. There is also potential for manufacturers to further reduce their recovery cost under individual responsibility models by designing their products to be more recyclable (see Atasu and Subramanian 2009). Thus, manufacturers should be aware of the type of legislation that works best for them to avoid fairness concerns.

Manufacturers' R&D and product design choices are also affected by take-back legislation. Plambeck and Wang (2009) show that especially in a competitive market, take-back legislation can improve the profitability of their organization by creating higher incentives to develop products of higher quality. This is because consumers are strategic and can anticipate the cost increase on the manufacturers through the EPR legislation, which in turn results in less-frequent product introductions, even under competition. Atasu and Subramanian (2009) make similar arguments by investigating the green design incentives coming from legislation and show that manufacturers can use the market valuation of recyclability to improve their profits. An important finding from their study is that the highest green design incentives would come from individual responsibility legislation. These are important messages for manufacturers with R&D and innovation as a core competency. Such organizations can benefit from certain types of legislation better than the others.

Interestingly, supply chain coordination can also help improve the efficiency of take-back legislation and manufacturer profits when facing such legislation. Given a specific form of legislation, manufacturers' supply chain coordination (see Subramanian et al. 2009) helps to improve not only the social welfare (see Jacobs and Subramanian 2009) but also supply chain profits and green design incentives in decentralized supply chains. Similarly, manufacturers can improve their profits by influencing the choice of their recovery channel partners. Toyasaki et al. (2008) show that competitive recycling markets favor manufacturers and consumers by increasing their profits.

In the end, all this discussion boils down to one critical point: the efficiency of manufacturing practices including supply chain choices, R&D decisions, and product design are directly tied to the form of take-back legislation faced. Manufacturers should find out how their core competencies/capabilities are affected by the specifics of legislation, be proactive, and seek the ultimate form of legislation implementation that will benefit them the most or harm them the least. For instance, a cost-efficient company should look for cost-reduction opportunities in such a legislation, while an innovative company should look for the possibility of individual action where the company can benefit from green design improvements. Social planners should also consider the discrepancies in the business environments and focus on improving the overall welfare that benefits the social welfare most. This can be done by giving up on a "one-size-fits-all" approach and developing alternative implementation possibilities for different categories of manufacturers. The legislation in Japan and the U.S. state of Maine seems to be in the right direction as they provide flexibility to the manufacturers to choose how they want to tackle the product take-back and recovery problem.

References

Atasu, A. and T. Boyaci. 2009. Take-Back Legislation and Its Impact on Closed Loop Supply Chains. *Working Paper*. Georgia Institute of Technology, Atlanta, GA.

Atasu, A. and R. Subramanian. 2009. Design Incentives in Take-Back Legislation. *Working Paper*. Georgia Institute of Technology, Atlanta, GA.

Atasu, A., L. N. Van Wassenhove, M. Dempsey, and C. Van Rossem. 2008. Developing Practical Approaches to Individual Producer Responsibility. *Working Paper*. INSEAD, Fontainebleau, France.

Atasu, A., M. Sarvary, and L. N. Van Wassenhove. 2009a. Efficient take-back legislation. *Production and Operations Management*, 18(3), 243–258.

Atasu, A., O. Ozdemir, and L. N. Van Wassenhove. 2009b. The Impact of Implementation Differences on the Efficiency of Take-Back Legislation. *Working Paper*. Georgia Institute of Technology, Atlanta, GA.

Calcott, P. and M. Walls. 2000. Can downstream waste disposal policies encourage upstream "design for environment"? *American Economic Review*, 90(2), 233–237.

Calcott, P. and M. Walls. 2002. Waste, Recycling, and Design for Environment: Roles for Markets and Policy Instruments, Resources for the Future. Discussion Paper 00-30REV.

Fullerton, D. and W. Wu. 1998. Policies for green design. *Journal of Environmental Economics and Management*, 36(2), 131–148.

Jacobs, B. and R. Subramanian. 2009. Sharing Responsibility for Product Recover Across the Supply Chain. *Working Paper*. Georgia Institute of Technology, Atlanta, GA.

Palmer, K., M. Walls, and H. Sigman. 1997. The cost of reducing municipal solid waste. *Journal of Environmental Economics and Management*, 33(2), 128–150.

Palmer, K. and M. Walls. 1997. Optimal policies for solid waste disposal taxes, subsidies, and standards. *Journal of Public Economics*, 65(2), 193–205.

Palmer, K. and M. Walls. 1999. Extended Product Responsibility: An Economic Assessment of Alternative Policies, Resources for the Future. Discussion Paper 99-12.

Plambeck, E.L. and Q. Wang. 2009. Effects of e-waste regulation on new product introduction. *Management Science*, 55(3), 333–347.

Subramanian, R., S. Gupta, and B. Talbot. 2009. Product design and supply coordination under extended producer responsibility. *Production and Operations Management*, 18(3), 259–277.

Toyasaki, F., T. Boyaci, and V. Verter. 2008. An Analysis of Monopolistic and Competitive Take-Back Schemes for WEE Recycling. *Working Paper*. McGill University, Montreal, Canada.

Walls, M. 2003. Extended Producer Responsibility and Product Design, Resources for the Future. Discussion Paper 03-11.

Walls, M. 2006. The Role of Economics in Extended Producer Responsibility: Making Policy Choices and Setting Policy Goals, Resources for the Future. Discussion Paper 06-08.

Walls, M. and K. Palmer. 2000. Upstream Pollution, Downstream Waste Disposal, and the Design of Comprehensive Environmental Policies, Resources for the Future. Discussion Paper 97-51-REV.

Chapter 4

Product Design Issues*

Bert Bras

Contents

4.1 Introduction

Product design for closed-loop supply systems is a relatively new field, but can draw upon decades of related experience from design for serviceability, maintainability, and (more recent) recyclability and remanufacturability. Although all these are relevant, a discussion of all design issues and guidelines would be beyond the scope (and text limit) of this chapter. This chapter, therefore, addresses only basic design issues primarily in the area of remanufacture and to some extent recycling, because these are the emerging areas of interest in the business community (the reader is encouraged to pursue further reading in the references cited). As it will become clear, product design is constrained by many factors—some known and some unknown at the time of design. Design for closed-loop supply chains is complicated by the fact that postconsumer returns can occur years after the product was designed in a world in which technology and business conditions have changed.

4.2 Design for re-X

Many authors postulate that a true closed-loop supply chain employs product remanufacture, but a "closed" closing supply chain has many options for what to do with products, and their embedded parts and materials. In fact, most of the time it would not make economic or environmental sense to pursue a 100 percent product remanufacturing (who would want a five-year-old cell phone?). Usually, products and their components undergo a combination of recovery, reuse, remanufacture, material recycling, reprocessing, incineration (for energy recovery), and disposal. Given this variety of options, we have coined the phrase "re-X" to capture the fact that design for closed-loop supply chains is not "just" a design for remanufacture or recycling issue. The "optimum" of this combination depends on economic as well as legislative factors that are often uncertain at the time of product design. Critical to successful product design is to know the process(es) you are designing for, and the critical technical and economic factors in these processes.

Design for re-X can be seen as a subset of the broad "design for X (DFX)" paradigm, but focused on end-of-service/life issues surrounding product design. From an economic and environmental value perspective, the most desirable and comprehensive re-X approaches are remanufacturing and recycling.

Remanufacturing is viewed differently from recycling in that the geometry of the product is maintained, whereas in recycling the product's materials are separated, shredded, ground, and molten for use in new product manufacture. Remanufacturing is viewed by many as a special form of recycling. The U.S. Code of Federal Regulations, for example, allows remanufactured products to be claimed as recyclable (see 16CFR 260.7), provided the conditions for such claims are met and conform to 16CFR260—"Guides for the use of environmental marketing claims." The German Engineering Standard VDI 2243 uses the phrase "product recycling" to denote product remanufacture in contrast to "material recycling" (VDI 1993). And the European End of Life Vehicle (ELV) Directive allows reuse to count as a form of recycling (EU 2000).

Although most will agree that remanufacturing typically offers the largest economic and environmental benefits, eventually a product and its parts will become obsolete. At that point, material recycling is often the preferred option, especially in light of legislative initiatives prohibiting the disposal and the incineration of certain products and materials. To understand the implications for product design, a discussion of some basic processes in these areas is warranted.

4.3 Remanufacturing Processes

4.3.1 Facility-Level Processes

To understand how to design for remanufacturing, one needs to know the basic processes. Remanufacturing spans many industry sectors and like in manufacturing no single uniform process exists. The following processes, however, can be found in any remanufacturing facility:

1. Warehousing of incoming cores, parts, and outgoing products
2. Sorting of incoming cores
3. Cleaning of cores
4. Disassembly of cores and subassemblies
5. Inspection of cores, subassemblies, and parts
6. Cleaning of specific parts and subassemblies
7. Parts repair or renewal
8. Testing of parts and subassemblies
9. Reassembly of parts, subassemblies, and products
10. Testing of subassemblies and finished products
11. Packaging
12. Shipping

More detailed discussions on some of these processes can be found in Bras (2007). Product design should be proper and focused on reducing the cost, effort, and overall resource expenditure of these processes. In general, however, deeper insight is needed, and a detailed study of actual processes may need to be performed similar to design for manufacturing efforts. To give a flavor of what the critical process issues are, in a survey (Hammond et al. 1998), a number of automotive remanufacturers provided insight into their most costly remanufacturing operations. Part replacement topped the list, followed by cleaning and refurbishing [see Hammond et al. (1998) for more survey results]. Although not exhaustive, such survey results are indicative of inherent product design problems.

4.3.2 Complicating Factors—External Actors

A common belief is that product design for closed-loop supply chains should make a product easier to (re)process for remanufacture, recycling, etc., at the end of its life. This is not necessarily true. In fact, market conditions may provide no incentives at all for designing products for reprocessing. And in many cases we start observing OEMs deliberately designing in features that attempt to prohibit remanufacturing. To understand this trend, a closer look at the different actors in closed-loop supply chains is warranted. Fundamentally, a number of different business practices exist with different combinations of actors. In Figure 4.1, a schematic of possible product flows is shown between different actors in the remanufacturing business practice. Two basic scenarios exist:

1. OEM manufactures, sells (or leases), recovers, remanufactures, and resells products and parts.
2. OEM manufactures and sells products, but third-party actors independently capture, remanufacture, and resell the used products and parts.

The second scenario is the predominant business scenario in many industries, whereas the first scenario is gaining momentum in certain industries. A hybrid scenario, that is, direct collaboration between OEMs and third-party remanufacturers, is also possible and frequently seen in automotive parts remanufacturing.

Clearly, there is no incentive for an OEM to improve its product design to facilitate full or partial product remanufacture if it does not benefit directly. In fact, most OEMs view independent remanufacturers as direct competition and will attempt to block remanufacturing by making, for example, disassembly difficult (e.g., by using sealed housings), using special information technology like chips in toner cartridges that have to be reset, etc. Even the widely known Kodak's single-use and funsaver cameras have evolved to become more difficult to disassemble

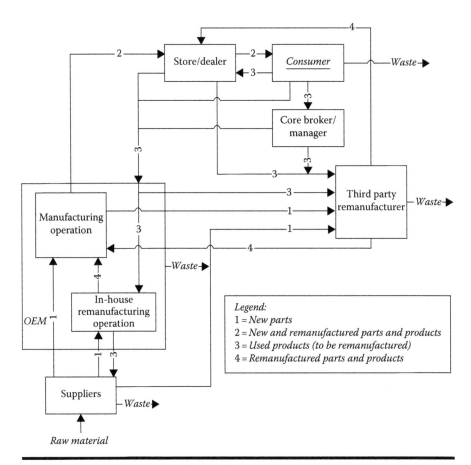

Figure 4.1 **Simple schematic of possible part and product flows in remanufacturing industry. (From Bras, B.,** *Handbook for Environmentally Conscious Mechanical Design***, Kutz, M. (Ed.), Wiley, New York, 2007, 283–318. With permission.)**

(without specialized tools) to counteract loss of returns and illegal film reloading by third-party photofinishers.

Similar design dynamics exist in designing products for material recycling. In Figure 4.2, a schematic of an automotive vehicle life cycle that illustrates these dynamics is given. Most of the material recycling is also done by third-party collectors/handlers and processors. There is no incentive for an OEM to design a product for the ease of recycling if others reap the benefits. Legislative initiatives like the European waste electrical and electronic equipment (WEEE) and ELV directives that put the ultimate responsibility on OEMs, however, have moved OEMs to collaborate with handlers and processors to comply with legislation and to reduce overall system cost.

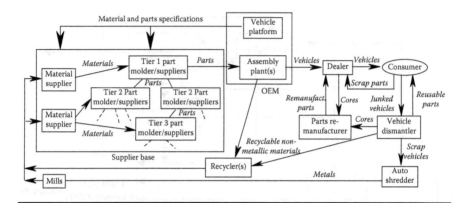

Figure 4.2 Schematic of automotive vehicle life cycle and industrial interest groups. (From Bras, B., *UN Ind. Environ.*, 20, 7, 1997. With permission.)

4.3.3 Other Factors Influencing Product Design

It should be clear from the preceding paragraph that there are many factors influencing product designs for remanufacture. In Figure 4.3, several factors are listed and divided into three classifications: individual product design characteristics; product development strategies and design management decisions; and business conditions and external factors. An in-depth overview of these factors can be found in McIntosh (1998) and McIntosh and Bras (1998a,b).

Basically, a designer at the product hardware level needs to be aware of the higher-level design strategies, and vice versa. Essentially, a designer's goal is to create a product that will be returned to the producer, has a large number of reusable components, and requires minimal disassembly, with required retrieval, disassembly, and remanufacture processes that are easy and inexpensive. But, as should be clear, the outcome can be very different depending on product development strategies and business conditions.

4.4 Overarching Design Principles and Strategies Enhancing Reuse

Prior to worrying about designing for facility-level remanufacturing processes, one should ensure that the actual product is even a candidate for reuse. Hence, one should first enhance the overall reusability of the product (or specific components) by proper design prior to worrying about specific remanufacturing process issues. In this context, market requirements are just as important as technical requirements. In this section, a number of overarching issues are highlighted that may enhance or hinder reuse and remanufacture.

Individual Product Design Characteristics:
1. Number of Components in Product Models
2. Design for Inexpensive and High Volume Product Retrieval
3. Number of Components Intended and Designed for Reuse
4. Level of Disassembly Required for Remanufacture
5. Ease of Component Disassembly
6. Ease of Remanufacture Processes (Sorting, Inspection, Cleaning, Refurbishment, Repair, Testing, Reassembly)

Product Development Strategies and Design Management Decisions:
1. Rate of Innovation (Technology Life Span of Components)
2. The Level of Product Variety (Number of Product Models at Any Time)
3. Developing Adaptable Products => Standardization Across Generations (Ability to Innovate While Preserving Components Across Generations)
4. Developing Families of Products => Standardization Across Product Variety (Ability to Offer Product Variety while Standardizing Across Product Models)
5. Volume of Each Product Model Produced
6. The Time Horizon Considered for Remanufacturing Assessments

Business Conditions and External Factors:
1. Cost to Reclaim Used Products
2. Percent of Used Products Returned
3. Product Life Span (Time Before Products are Returned)
4. Shifts in Technology and Consumer Requirements (Changes in Technology Life Span Beyond Designer's Control)
5. Cost of Manufacturing New Components
6. Prices Received for Scrap Recycling of Components
7. Cost of Labor and other Remanufacturing Process Costs
8. Remanufacturing Process Setup Costs
9. Inefficiency When New Remanufacturing Processes Are Required
10. Government Policies (Incentives for Reuse or Recycling)
11. Producer's Opportunity Cost of Capital

Figure 4.3 Decisions and factors influencing remanufacturing viability. (From McIntosh, M.W. and Bras, B.A., *1998 ASME Design for Manufacture Conference, ASME Design Technical Conferences and Computers in Engineering Conference,* Atlanta, GA, 1998b.)

4.4.1 Product or Component Remanufacture?

Remanufacturing should be part of a larger business strategy. As such, products should not be designed "just" for remanufacturing, but also for functionality, initial manufacturability, etc. Depending on the situation, conflicts with other design guidelines can occur and detailed design analyses may need to be performed.

When designing a product, it should be kept in mind that remanufacturing the entire product may not be the best strategy and is often more an exception than the rule. Rather, the remanufacture of certain product subassemblies is often more appropriate. A rather trivial example of this is an automobile. Power train components are commonly remanufactured, but interiors and bodies are not.

Similarly, remanufacturing entire products can be bad for the environment. Consider the fact if appliances and automobiles from the 1950s were kept in service as is through remanufacturing. We would have much higher energy consumption due to their older and more inefficient technology. Clearly, remanufacturing has its limitations. Leading OEMs who have internalized remanufacturing as part of their business, therefore, will spend significant time designing a product architecture that allows for technology upgrades. Fuji-Xerox, for example, looked five years ahead to see what technology may need to be incorporated in its copier systems and identifies which components should be designed as replaceable by upgrades versus which components should be designed for reuse (Gutowski et al. 2001; Allen et al. 2002). Also in manufacturing equipment, we see that such "upward" remanufacturing is done by adding new control systems. Hence, 100 percent reuse of all components is typically not feasible or even desirable. Finding what to reuse and what to replace by upgrades and how to design the architecture around that is the first major challenge for OEM designers.

4.4.2 Product Architecture Design Guidelines

Products become obsolete and are replaced because of

1. Degraded performance, including structural fatigue, caused by normal wear over repeated uses
2. Environmental or chemical degradation of (internal) components
3. Damage caused by accident or inappropriate use
4. Newer technology becomes available prompting product replacement
5. Fashion changes

In general, the first three categories tend to be driving product returns and remanufacturing of mechanical engineering products. The replacement of information technology products (e.g., computers) is mostly caused by rapid technology changes. Consumer electronic products (e.g., cell phones) are examples where products are simply being replaced due to newer technology or changes in fashion. The replaced products are often fully functional and well within their operating specifications. In such cases, the remanufacturing process may collapse to a simple "collect, test, and resell or discard" operation.

To achieve a high degree of product or component reuse, the above-mentioned causes for obsolescence have to be countered. Components and subassemblies that are good candidates for reuse, therefore, have the following characteristics:

■ Stable technology (not much change expected in the product's lifetime)
■ Damage resistant
■ Aesthetics and fashion are (largely) irrelevant

Given that we often do not exactly know future technology or fashion demands, a critical issue is therefore the "openness" of the product design to future modifications and upgrades. Upgradeable products allow for a larger percentage to be salvaged.

Strive for open systems and platform designs that have modular product structures to avoid technical obsolescence. Platform design attempts to reduce component count by standardizing components and subassemblies while maximizing product diversity. Designing the product in modules allows the upgradation of function and performance (e.g., computers) and the replacement of technically or aesthetically outdated modules (e.g., furniture covers). As mentioned before, Fuji-Xerox develops multiyear upgrade plans and associated product modules for its copier design. More information on modular design can be found in Newcomb et al. (1998) where a method is described to design products with consistent modularity with respect to life-cycle viewpoints such as servicing and recycling. The authors define modularity with respect to life-cycle concerns in addition to modularity just meaning a correspondence between form and function.

Strive for a "classic" design to avoid fashion obsolescence. Aesthetically appealing and "timeless" designs are usually more desirable (higher priced), better maintained, and have greater potential for long life spans and multiple reuse cycles. This is more in the realm of industrial design than mechanical design, but designing a product that does not become uninteresting or unpleasing quicker than its technical life will reduce the product's obsolescence and increase its desirability and potential for reuse.

Strive for damage-resistant designs. Although this sounds like basic good engineering, lighter duty materials and smaller, more optimized, part sizes and geometries are engineering design aspects that potentially reduce the number of service cycles, and can become problems in various facets of remanufacturing. Both are directly related to design, as current designs are being optimized primarily to reduce weight, space, and cost. A good example is the reduction in wall thicknesses between cylinders in engine blocks. This reduces mass, but it also affects remanufacturability because damage due to, for example, scoring in the cylinder walls cannot be removed using machining. Instead a sleeve may have to be inserted, but this may not be possible due to the thin walls. Clearly, this practice benefits the manufacturer, but can cause difficulty for remanufacturers. Figure 4.4 shows a clutch pressure plate that has a broken ear. Rough handling (e.g., by dropping it) in shipping or removal may have caused this failure. The plate (cast iron with machined surfaces) can only be salvaged using (expensive) welding and testing processes. This type of accidental failure will only increase if parts are designed closer to strength and endurance limits.

4.4.3 Product Maintenance and Repair Guidelines

The service life of products can be extended in two basic ways: (1) make the product stronger and more durable and (2) allow for good maintenance. Overall reliability

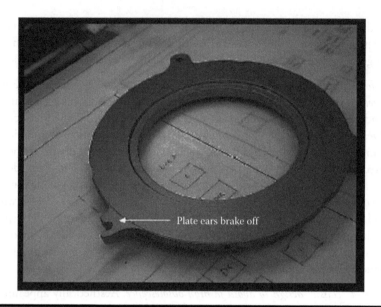

Plate ears brake off

Figure 4.4 Damage to clutch pressure plate ear.

and durability is enhanced by following solid engineering principles in developing a sound design and avoiding weak links. Methods such as failure mode and effect analysis are effective approaches to check the design.

Although maintenance is needed for many products, incorrect maintenance can have often disastrous effects. For example, car owners may add the wrong type of oil to their automotive engines and transmissions. The design team can choose whether (a) to allow for (user) maintenance and run the risk of unintended failures due to poor maintenance or (b) to design the product so that it is either maintenance free or can only be maintained through specialized (OEM) personnel. In general, the latter is preferable when it is known that non-qualified personnel (such as users) will attempt prescribed maintenance operations. Maintenance by OEM personnel also adds a new business dimension for the OEM's overall business strategy.

Given that qualified personnel are available for maintenance, designs should allow for easy maintenance and repair where needed. Product design should follow available design for serviceability guidelines. Again, the best strategy, typically, is to design the product such that it needs little or no maintenance, or only maintenance by expert personnel. If maintenance has to be done by users, it should be designed absolutely foolproof. Some strategies for achieving easy maintenance are as follows:

- Indicate on the product how it should be opened for cleaning or repair
- Indicate on the product itself which parts must be cleaned or maintained, for example, by color-coding lubricating points

- Indicate on the product which parts or subassemblies are to be inspected often due to rapid wear
- Make the location of wear detectable so that repair or replacement can take place on time
- Locate the parts that wear relatively quickly close to one another and within easy reach
- Make the most vulnerable components easy to dismantle

Plus, provide clear maintenance and repair manuals and communication. Consider including vital information on the actual product itself too. Good examples are stickers or labels with tire pressure ratings and oil-type requirements placed in cars, which also aids service personnel.

4.4.4 Design for Reverse Logistics

If the remanufacturing process is part of an OEM's integrated strategy, core collection and reverse logistics also become crucial processes that can be aided by design. Core collection can be done by independent core managers or core brokers, through third-party subsidiaries or suppliers/customers (e.g., single-use cameras through photofinishers and automotive parts through parts stores), or through direct channels (e.g., direct mail in of toner cartridges to OEMs). Although often overlooked, the design of easy to use and protective single or bulk packaging can greatly increase core returns. Good examples are toner cartridges that come in returnable boxes with prearranged return addresses and shipping labels.

4.4.5 Parts Proliferation versus Standardization

Product diversity (or "part proliferation") is a significant problem especially in automotive parts remanufacturing. In automotive remanufacturing, the term "part proliferation" refers to the practice of making many variations of the same product—differing only in one or two minor areas. However, these differences (such as electrical connectors) are distinct enough to prevent interchanging these similar products. For example, for a given model year, a car line may have one or more different alternators for each variation of the vehicle—the alternator for the two-door model would not be able to be used to replace the alternator for the four-door model. Not only can they not be used within the car line, but no other car line made by the manufacturer can use the part either. To exemplify the amount of parts proliferation in the 1980s, consider the following numbers from an Atlanta-based large automotive remanufacturer. In 1983, there were approximately 3,400 different part numbers for brake products whereas by 1995, there were approximately 16,500 different part numbers!

Problems arising from this practice range from having to keep a large inventory of replacement parts, to having to keep track of several, non-standardized assembly and disassembly processes. An increase in the variety of assembly and disassembly processes also results in an increase in the number of process setups that have to be made, causing a reduction in throughput. Employee training also becomes a significant issue as a result, as they must be familiarized with all of the various, unique parts and the processes for each new product.

It is interesting to note that the trend of parts proliferation in the automotive sector started in the early 1980s. Among others, this coincides with the move of major U.S. automakers to a platform organization and a move toward lean production. Between 1982 and 1990, Japanese automakers nearly doubled the number of models on the road, from 47 to 84 models. Reacting to this condition, U.S. automakers also increased their models on the road from 36 to 53 in the same period of time (Womack et al. 1991). Furthermore, the independence of individual platforms within an automaker's organization seems to have led to a reduction of shared components among automotive models, resulting in decreased standardization and increased parts proliferation.

A good design practice to counterpart proliferation is to design products using standard parts. Standardization always supports remanufacture, and also manufacturing operations, and should be pursued wherever possible. Among others, standardization reduces the number of different tools needed to assemble and disassemble and increases economies of scale in replacement part purchasing, eases warehousing, etc. Different product aspects can be standardized:

- *Components*: Use as much as possible standard, commonly and easily available components. Use of specialty components may render the remanufacture of assemblies impossible if these specialty components cannot be obtained any more.
- *Fasteners*: By standardizing the fasteners to be used in parts, the number of different fasteners can be reduced, thus reducing the complexity of assembly and disassembly, as well as the material-handling processes.
- *Interfaces*: By standardizing the interfaces of components, fewer parts are needed to produce a large variety of similar products. This helps to build economies of scale, which also improves remanufacturability. The PCI interface standard in computers is a good example of a standard interface.
- *Tools*: Ensure that the part can be remanufactured using commonly available tools. The use of specialty tools can also degrade serviceability.

4.4.6 Hazardous Materials and Substances of Concern

A critical issue is to avoid hazardous substances and materials of concern. Products that contain hazardous materials (a) require specialized processing equipment (higher capital costs) and (b) will be in lower demand, resulting in low(er) profit margins. Plastics that contain halogenated flame retardants are a good example of this in the material recycling domain. Although a large volume of these exist suitable for recycling, recyclers cannot find markets for these plastics. Sometimes, hazardous materials can be removed and retrofitted using nonhazardous materials during remanufacturing. Air-conditioning and refrigeration systems that used Freon are examples where a new refrigerant can be substituted. Performance, however, may degrade slightly because the product design was not necessarily optimized for the new refrigerant. Regardless of the ability to retrofit, one should always strive to reduce the number of parts that contain environmentally hazardous materials. Also, machining or, otherwise, processing of parts with (heavy) metals like chromium, zinc, lead, etc., may trigger EPA toxic release inventory (TRI) reporting and require special air-handling equipment as per federal and local regulations, adding to remanufacturing costs.

4.4.7 Intentional Use of Proprietary Technology

The use of technology that is proprietary or difficult to reverse engineer will block/limit the number of independent entrepreneurs remanufacturing OEM parts and products. This practice has started to emerge as certain OEMs have realized the value of remanufactured products and how third-party remanufacturers can take away the market share of OEM product and component sales. The inkjet printing industry has several examples where an OEM has included chips that can only be reset by an authorized remanufacturer. Similarly, Kodak's single-use cameras became more difficult to disassemble with common available tools to counter third-party film reloading and reuse. This strategy is counter to what many academics say what should be done regarding product design for remanufacture, but this practice clearly makes sense from a higher-level business strategy where an OEM wants to retain market share and sales.

4.4.8 Inherent Uncertainties

Last but not least, in remanufacture, the number and the range of uncertainties are higher than that for "regular" manufacture and logistics because many of the concerns are out of the control of the OEM and the designers. Some sample product uncertainties encountered are as follows:

- How long is a typical use or life span?
- What is its state after its each use?
- What changes have been made during use and throughout its life?

This affects organizational uncertainties such as

- How many will be available for take-back, and when?
- How long will it take to reprocess the product?
- What is the demand?

Some remanufacturing operations have throughput yields as low as 40–60 percent (unheard of in manufacturing) due to a combination of poor-quality cores and poor processing. Designers and product realization teams should be aware of these uncertainties, and ideally try to manage or even eliminate the uncertainties by smart product and process design. For example, changes can be avoided if the product design eliminates the possibility of user tampering.

4.5 Hardware Design Guidelines

In the preceding section, some specific design guidelines were given that enhance the overall suitability of remanufacturing a given product. In this section, some specific component and machine design-type guidelines are given that primarily facilitate the facility-level remanufacturing processes. Clearly, this discussion is not exhaustive and the reader is encouraged to use his or her own engineering insight as well to identify design guidelines for his or her own remanufacturing operations and product designs.

4.5.1 Basic Sources and Overviews

There are relatively few publications and sources with general design for remanufacturing guidelines in existence. The emergence of WEEE and ELV take-back directives from the European Union (EU 2000, 2003), however, has resulted in a number of design-for-recycling guidelines—some of which are applicable to remanufacturing. General design-for-recycling guidelines were formalized in the German Engineering Standard, VDI 2243 (1993). These guidelines also contain directional criteria for the design of remanufacturable products. According to VDI 2243 and other sources (Lund 1984; Haynsworth and Lyons 1987; U.S. Congress 1992; Beitz 1993; Berko-Boateng et al. 1993), remanufacturable assemblies should be designed with special emphasis on the following:

- *Ease of disassembly*: Where disassembly cannot be bypassed, by making it easier, less time can be spent during this non-value-added phase. Permanent fastening such as welding or crimping should not be used if the product is intended for remanufacture. Also, it is important that no part be damaged by the removal of another.
- *Ease of cleaning*: Parts that have seen use inevitably need to be cleaned. To design parts such that they may easily be cleaned, the designer must know what cleaning methods may be used, and design the parts such that the surfaces to be cleaned are accessible, and will not collect residue from cleaning (detergents, abrasives, ash, etc.).
- *Ease of inspection*: As with disassembly, inspection is an important, yet a non-value-added, phase. The time that must be spent on this phase should be minimized.

- *Ease of part replacement*: It is important that parts that wear are capable of being replaced easily, not just to minimize the time required to reassemble the product, but to prevent damage during part insertion.
- *Ease of reassembly*: As with the previous criteria, time spent on reassembly should be minimized using design for assembly guidelines (Boothroyd and Dewhurst 1991). Where remanufactured product is assembled more than once, this is very important. Tolerances also relate to reassembly issues.
- *Reusable components*: As more parts in a product can be reused, it becomes more cost effective to remanufacture the product (especially if these parts are costly to replace).

In the following section, we focus on a number of guidelines in more detail. Clearly, the inherent and underlying assumption is that the products are being designed for remanufacture by an OEM or a "friendly" third party. Otherwise, there is no incentive to follow any of these design guidelines.

4.5.2 Sorting Guidelines

Sorting is the first step in any remanufacturing process. Mostly, it is coupled with an initial inspection as well. Figure 4.5 is illustrative of how cores are received by many third-party remanufacturers.

The container in Figure 4.5 contains boxed and unboxed starters, alternators, and brake shoes of varying types, shapes, sizes, and conditions. In such cases,

Figure 4.5   Cores arrive at automotive remanufacturer.

worker knowledge and expertise are key in the sorting process. Product and part design can facilitate the sorting process by following some guidelines:

- *Reduce product and part variety.* The less different parts need to be sorted, the less time it costs.* This also implies for internal components. The standardization of fasteners, bearings, pulleys, etc., will greatly speed up initial core as well as subsequent part sorting.
- *Provide clear distinctive features that allow for easy recognition.* If different parts have to be used, make sure that they are easily recognizable. For example, having two housings being exactly the same except for one different-sized hole may not be the best strategy because the sorter/inspector has to distinguish bases on small size differences. A binary yes/no-type distinction is much easier to do and can be achieved by, for example, changing the hole pattern.
- *Provide (machine) readable labels, text, bar codes that do not wear off during the product's service life.* Most products and parts have labels. Those that are exposed to the environment, however, tend to wear off during life unless they have been stamped, casted, or molded in. Even riveted serial plates and numbers can shear and wear off. Internal parts fair better provided they have part numbers. Some companies are experimenting with radio-frequency identification (RFID) tags to facilitate sorting, but that is rather the exception than the rule.

4.5.3 Disassembly Guidelines

A phrase often heard is "If a remanufacturer can take a product apart, it can be remanufactured." At first, this statement would seem to indicate that the design should focus on disassembly to ensure that the product can be remanufactured. However, there is a hidden assumption in this statement. A more correct statement is "If a remanufacturer can take a product apart without damaging important parts, it can be remanufactured." The two key ideas that designers should extract from this statement are nondestructive disassembly and preventing key parts from being damaged.

In remanufacture, the objective is to reuse cores and components. This means that (in contrast to material recycling) destructive disassembly techniques like shredding are not an option. Manual disassembly, supported by pneumatic or other handheld mechanized means, is the general norm of the industry—for better or for worse. Proper design can make disassembly easier so that less time can be spent during this non-value-added phase, but the goal of remanufacture is to salvage cores and components of value, and any damage must be repaired. Speedy

---

* This can also be achieved by remanufacturers themselves through specialization on specific products and cores.

disassembly is desired, but not at the expense of damaging cores. Avoiding and preventing damage, therefore, is often the more important objective than increasing speed. Given this, we can define a number of simple overarching guidelines for fasteners.

4.5.3.1 Avoid and Prevent Damage

- Avoid permanent fasteners that require destructive removal (such as rivets, welds, and crimp joints).
- If fasteners require destructive removal, ensure that their removal will not result in damage to core and other reusable parts by incorporating break-points or appropriate strong lever points.
- Reduce the number of fasteners prone to damage and breakage during removal (e.g., snap fits). For example, Phillips/Blade/Torx fasteners are more easily prone to head damage and removal difficulties than hex and Allen bolts. Molded plastic snap fits often break due to the aging of the plastic, either causing a need for repair or resulting in the whole part to be scrapped.
- Increase the corrosion resistance of fasteners, where appropriate. This reduces damage and facilitates removal.

4.5.3.2 Increase Speed

- Reduce the total number of fasteners in a unit.
- Reduce the number of press-fits that do not have "push-out" capability.
- Reduce the number of fasteners without direct line of sight.
- Standardize fasteners by reducing the number of different types of fastener (Hex/Phillips/Allen/Torx, metric/SAE, etc.). Reducing the number of different size fasteners (i.e., length, diameter) will speed up reassembly and allow for larger economies of scale in purchasing fasteners.

4.5.4 Design for Reassembly

Reassembly, the last process in a typical remanufacturing process, is basically identical to assembly in manufacturing. To design for reassembly, follow common design for assembly guidelines. Table 4.1 contains common design for assembly guidelines that can be found in the general literature.

Manufacturers tend to use design for assembly and manufacturing processes that make it difficult for parts to be reused or remanufactured. For example, solenoids for starter motors are crimped into their housings. Not only is it difficult to remove the crimps to remanufacture the solenoid, but also crimped fasteners cannot be re-crimped without degrading the strength of the crimp.

Table 4.1 Common Design for Assembly Guidelines

1.	Overall component count should be minimized.
2.	Minimize use of fasteners.
3.	Design the product with a base for locating other components.
4.	Do not require the base to be repositioned during assembly.
5.	Design components to mate through straight-line assembly, all from the same direction.
6.	Maximize component accessibility.
7.	Make the assembly sequence efficient.
	• Assemble with the fewest steps.
	• Avoid risks of damaging components.
	• Avoid awkward and unstable component, equipment, and personnel positions.
	• Avoid creating many disconnected subassemblies to be joined later.
8.	Avoid component characteristics that complicate retrieval (tangling, nesting, and flexibility).
9.	Design components for a specific type of retrieval, handling, and insertion.
10.	Design components for end-to-end symmetry when possible.
11.	Design components for symmetry about their axes of insertion.
12.	Design components that are not symmetric about their axes of insertion to be clearly asymmetric.
13.	Make use of chamfers, leads, and compliance to facilitate insertion.

4.5.5 Cleaning

Parts that have seen use inevitably need to be cleaned. To design parts such that they may easily be cleaned, the designer must know what cleaning methods may be used, and design the parts such that the surfaces to be cleaned are accessible, and will not collect residue from cleaning (detergents, abrasives, ash, etc.). The following guidelines capture the basic aspects:

■ *Protect parts and surfaces against corrosion and dirt.* The best strategy is to minimize cleaning wherever and whenever. Proper corrosion coating and dirt protection will support this. However, also consider that any coating (e.g., paint) may have to be removed if damaged. Hence, a balance may have to be found between protection and ease of removal.

■ *Avoid product or part features that can be damaged during cleaning processes, or make them removable.* For example, when thermal cleaning is used, make sure all materials can withstand the heat without adverse effects. Abrasive cleaning methods can gouge surfaces.

■ *Minimize geometric features that trap contaminants over the service life.* A sharp concave corner is an example of a geometric feature that traps contaminants. If a rib or plate is expected to trap dirt or grease, consider making it removable.

■ *Reduce the number of cavities/orifices that are capable of collecting residue (abrasives, chemicals, etc.) during cleaning operations.* Any orifice that can collect dirt or cleaning debris will have to be plugged or cleaned afterward.

■ *Avoid contamination caused by wear. Internal components can become "dirty" due to wear of other components.* For example, oil seals may wear and the resulting leakage will cause the contamination of other parts. Proper shielding or designing of such sources of wear can reduce the cleaning effort required.

4.5.6 Replacement, Reconditioning, Repair

In general, remanufacturing tends to avoid the replacement of parts, but there are trade-offs as to whether to spend money to buy a new part or spend money to repair the part. For commonly available parts like bearings and fasteners, the choice is easy, but the higher the part price, the more incentive for refurbishment instead of replacement. The cost of replacement can be reduced by the following guidelines:

■ Reduce the number of parts subject to wear
■ Avoid materials that degrade through corrosion
■ Reduce the number of parts to be removed to gain access to damaged parts to be replaced (or refurbished)
■ Reduce the number of independently functioning parts that are inseparably coupled
■ Reduce the number of special parts (including aesthetic features)

As discussed in Section 4.3, there are a number of basic strategies for repairing damage and refurbishing surfaces. Proper material selection can aid remanufacturing, as well as surface protection. An interesting problem with surface protection like heavy-duty paint or powder coating is that it protects a part, but can cause significant cleaning problems in remanufacturing when the coating needs to be removed for renewal. For surface reconditioning like painting, plating, etc., consider the following guidelines:

- Reduce the number of parts whose surface finish cannot be refinished through commonly available and conventional means
- Minimize the number of orifices that must be masked prior to painting
- Reduce the number of (exterior) parts that must be removed prior to painting

Also, minimize the number of parts that can retain dents/deformations.

4.5.7 Inspection and Testing

Inspection and testing can be facilitated by reducing the number of different testing and inspection equipment pieces needed, as well as by reducing the level of sophistication required. Although not in the realm of product design per se, good testing documentation and specifications should be provided to ensure that the correct specifications are achieved and tested for. This assumes (again) OEM involvement in the remanufacturing process.

4.6 Product Design for Material Recycling

Recycling is often defined as a series of activities, including collection, separation, and processing, by which products or other materials are recovered from or otherwise diverted from the solid waste stream for use in the form of raw materials in the manufacture of new products. In essence, one can argue that any product design for a closed-loop supply chain should facilitate material recycling because, eventually, all products and parts will become obsolete. The emergence of WEEE and ELV take-back directives from the European Union (EU 2000, 2003) has resulted in a number of design-for-recycling guidelines and the field. A good overview of general design-for-recycling guidelines is formalized in the German Engineering Standard, VDI 2243 (VDI 1993).

To properly design a product for recycling, one should (again) know the processes involved in such recycling operations. As mentioned, typical recycling processes include a combination of collection, sorting, storage, manual separation of assemblies, various stages of mechanical separation of materials (dependent on desired material purity), reprocessing of materials. Often one can distinguish actors in recycling between collectors/handlers and processors. A collector/handler focuses on gathering, sorting, and some preliminary (manual) separation of products, subassemblies, and materials. These are then sold and shipped to more specialized processors who can efficiently and effectively separate, purify, and reprocess the materials for use in new products through a variety of mechanical, thermal, and chemical processes. Due to the higher capital investment, fewer processors exist other than collectors/handlers. Some processors operate in conjunction with large material producers (or are one and the same); others operate independently and sell their material on the free market. Examples of such processors are printed circuit board (PCB) and cathode ray tube (CRT) glass processors that receive their input stream from electronic waste collectors who, for example, separate the PCBs and CRTs from desktop computers. Computer housing materials are shredded and shipped to plastics or metal processors. From a design point of view, it is important to understand that modern recycling relies on mechanized processes—augmented with human labor where necessary or economically preferable. Key issues that have been learned over the years are (Coulter et al. 1998) as follows:

- The limiting factor in economic recycling of complex, integrated assemblies is the separation into pure material streams.
- Both manual and mechanical separations have their advantages and disadvantages.
- Significant value must be retained in a part for manual separation to be economically viable.
- Different design techniques should be employed depending on whether one wants to facilitate manual separation or mechanical separation.

Manual and mechanical separations have different requirements. Mechanized separation techniques exploit and rely on differences in material properties. For example, automotive recycling exploits the magnetic property of steel to separate steel (using magnets) from other nonmagnetic materials. Entire vehicles are shredded in fist-size particles, which are then separated on conveyor belts using magnets and Eddy current separators.

Design for mechanical separation, therefore, requires more effort in creating an assembly or component that can be separated quickly and easily into pure streams of materials based on material properties. If the materials used do not have distinctive properties that can be used for separation, economical recycling will not be possible. Disassembly effort and visual identification are not important for mechanical separation, but material selection is critical to this separation effort.

Manual separation requires more effort on improving the disassembly and sorting process for the component or assembly, because the primary limiting factor for manual separation is the (labor) time required for this separation. If the materials in the part being considered require significant time to separate and identify, manual separation will not be economically feasible. In general, manual disassembly is preferred when the nondestructive removal is required of

- Large amounts of materials with high purity
- Parts with regulated materials that could contaminate a material stream (like batteries PCBs, lead glass, etc.) and require separate handling
- Parts destined for remanufacture

Guidelines appropriate for manual separation are those that suggest ways to reduce the (manual) disassembly effort through appropriate fastener selection, the avoidance of obstructions, and facilitating visual material identification through markings.

Many design guidelines—such as the reduction of the number of materials used, the standardization of material types, and the use of recyclable materials—are applicable for either type of separation. There are, however, a significant number of possible design techniques that are only useful for one type of separation. The distinction in designing for manual or mechanical separation is illustrated with fastener selection, material selection, and component design guidelines.

Fastener selection: A number of different techniques can be used when designing for manual separation, all intended to reduce the amount of time it takes to dismantle the components. Specifically, these include reducing the number of fasteners, communizing the fastener types, using snap fits, and avoiding non-removable fasteners. When considering mechanical disassembly, the only concern is the separability of the fastener material from other materials in the component because the fasteners will be shredded with the component. Accordingly, the number and type of fastener used is not important. Instead, integral fasteners and material-compatible fasteners are greatly preferred. If this is not possible, ferrous fasteners are preferred in plastic assemblies because they allow for easy magnetic separation.

Material selection: Perhaps the most interesting distinction in the material selection guidelines is that between component and assembly. If an assembly that contains two polymers is being considered for manual separation, the designer should attempt to create components made of one material or the other, so that the components do not need to be disassembled as well. However, if the same assembly is being designed for mechanical separation, it does not matter whether the individual components of the assembly are mixed materials or not as the entire assembly will be shredded anyway. For manual separation, large masses of a single material are important. For mechanical separation, reducing the total number of different materials in the assembly is more important. In addition, it is extremely important to note the specific material properties that will be used in the mechanical separation and to make sure that there is sufficient distinction to allow easy and accurate separation. A metal plate riveted to a plastic component would be extremely difficult to disassemble manually, yet can easily be separated using mechanical means.

Component design: For manual disassembly, a number of techniques are useful for decreasing disassembly time. One of these is simply the application of design for serviceability guidelines, because a component that is easy to disassemble for servicing will usually be easy to disassemble for recycling. Although this relationship tends to fall apart for small components with mixed materials (as seen in the disassembly of the luxury sedan), it still provides a benefit to both serviceability and recyclability. For mechanical separation, of course, designing for serviceability does not affect recyclability. In fact, components that must be serviced are strong candidates for manual separation as they must be easy to disassemble.

4.7 Design for Manufacturing Conflicts

It should be noted that, in some cases, design for remanufacturing can conflict design for manufacturing, and even be not in the best interest for the environment. For example, increasing longevity by adding material can increase part weight, causing more upfront material expenditures (and cost) and potentially more fuel consumption and emissions in transportation systems. Some differences also exist between design for disassembly versus design for assembly. For example, complete nesting can slow disassembly by not providing a location for the disassembler to reach, grasp, or otherwise handle (Noller 1992). As noted in Scheuring et al. (1994a,b), the main negative effects on assembly for the most part deal with making easily separable joints. This would negatively affect assembly in the sense that the purpose of the assembly step could be easily negated during product use. A compromise solution would be to design joints that are very hard to disassemble during product use but easy to dismantle after the customer use or for the purpose of servicing a product. Different design for disassembly strategies will have different effects on the overall processing time, especially if coupled with reassembly processes. In Scheuring (1994) and Scheuring et al. (1994a,b), a study on single-use cameras suggested that a modular design was slightly more effective in improving disassembly efficiency than parts consolidation, and much more effective than reducing orientation changes during disassembly. Clearly, as indicated earlier, good design should take a life-cycle perspective—both from economic and environmental points of view.

4.8 Conclusion

In this chapter, various product design issues related to closed-loop supply chains with special emphasis on remanufacturing and recycling were discussed. A distinction was made between overarching guidelines versus specific component hardware-oriented design guidelines. As was shown, a solid understanding of the re-X processes employed is critical to performing a good product design. Furthermore, unless an OEM is benefiting, there is little incentive for an OEM (or its suppliers) to design products for remanufacture, recycling, or any other re-X activity.

References

Allen, D. T., D. J. Bauer, B. Bras, T. G. Gutowski, C. F. Murphy, T. S. Piwonka, P. S. Sheng, J. W. Sutherland, D. L. Thurston, and E. E. Wolff (2002). Environmentally benign manufacturing: Trends in Europe, Japan and the USA. *ASME Journal of Manufacturing Science* **124**(4): 908–920.

Beitz, W. (1993). Designing for ease of recycling—General approach and industrial applications. In *Ninth International Conference on Engineering Design*, the Hague, the Netherlands. Zurich, Switzerland: Heurista.

Berko-Boateng, V. J., J. Azar, E. De Jong, and G. A. Yander (1993). Asset recycle management—A total approach to product design for the environment. In *International Symposium on Electronics and the Environment*, Arlington, VA, IEEE.

Boothroyd, G. and P. Dewhurst (1991). *Product Design for Assembly*. Wakefield, MA: Boothroyd and Dewhurst, Inc.

Bras, B. (1997). Incorporating environmental issues in product realization. *United Nations Industry and Environment* **20**(1–2): 7–13.

Bras, B. (2007). Chapter 8 – Design for remanufacturing processes. In *Handbook for Environmentally Conscious Mechanical Design*, M. Kutz (Ed.). New York: Wiley, pp. 283–318.

Coulter, S. L., B. A. Bras, G. Winslow, and S. Yester (1998). Designing for material separation: Lessons from the automotive recycling. *Journal of Mechanical Design* **120**(3): 501–509.

EU (2000). Directive 2000/53/EC of the European Parliament and of the Council of September 18 2000 on End-Of Life Vehicles. *Official Journal of the European Communities* L 269:34–42.

EU (2003). Directive 2002/96/EC of the European Parliament and of the Council of 27 January 2003 on Waste Electrical and Electronic Equipment. *Official Journal of the European Communities* L 37:24–38.

Gutowski, T. G., C. F. Murphy, D. T. Allen, D. J. Bauer, B. Bras, T. S. Piwonka, P. S. Sheng, J. W. Sutherland, D. L. Thurston, and E. E. Wolff (2001). *Environmentally Benign Manufacturing*. Baltimore, MD: World Technology (WTEC) Division, International Technology Research Institute.

Hammond, R., T. Amezquita, and B. Bras (1998). Issues in automotive parts remanufacturing industry: Discussion of results from surveys performed among remanufacturers. *Journal of Engineering Design and Automation, Special Issue on Environmentally Conscious Design and Manufacturing* **4**(1): 27–46.

Haynsworth, H. C. and R. T. Lyons (1987). Remanufacturing by design, the missing link. *Production and Inventory Management* Second Quarter: 25–28.

Lund, R. T. (1984). Remanufacturing. *Technology Review* **87**: 18–23.

McIntosh, M. W. (1998). Modeling the value of remanufacture in an integrated manufacture-remanufacture organization. MS thesis, George W. Woodruff School of Mechanical Engineering, Georgia Institute of Technology, Atlanta, GA.

McIntosh, M. W. and B. A. Bras (1998a). Addressing rapid innovation and mass customization in an integrated manufacturing-remanufacturing organization. In *Fifth International Congress on Environmentally Conscious Design and Manufacture (Regenerative Design: The New Millennium)*, Rochester, NY.

McIntosh, M. W. and B. A. Bras (1998b). Determining the value of remanufacture in an integrated manufacturing-remanufacturing organization. In *1998 ASME Design for Manufacture Conference, ASME Design Technical Conferences and Computers in Engineering Conference*, Atlanta, GA, ASME.

Newcomb, P. J., B. A. Bras, and D. W. Rosen (1998). Implications of modularity on product design for the life cycle. *Journal of Mechanical Design* **120**(3): 483–490.

Noller, R. M. (1992). Design for disassembly tactics. *Assembly* January: 24–26.

Scheuring, J. F. (1994). Product design for disassembly. MS thesis, George W. Woodruff School of Mechanical Engineering, Georgia Institute of Technology, Atlanta, GA.

Scheuring, J. F., B. A. Bras, and K.-M. Lee (1994a). Effects of design for disassembly on integrated disassembly and assembly processes. In *Fourth International Conference on Computer Integrated Manufacturing and Automation Technology*, Rensselaer Polytechnic Institute, Troy, NY, IEEE.

Scheuring, J. F., B. A. Bras, and K.-M. Lee (1994b). Significance of design for disassembly in integrated disassembly and assembly processes. *International Journal of Environmentally Conscious Design and Manufacturing* **3**(2): 21–33.

U.S. Congress (1992). *Green Products by Design: Choices for a Cleaner Environment*. Washington, DC: U. S. Congress, Office of Technology Assessment.

VDI (1993). VDI 2243—Konstruieren Recyclinggerechter Technischer Produkte (Designing Technical Products for ease of Recycling). VDI-Gesellschaft Entwicklung Konstruktion Vertrieb, Germany.

Womack, J. P., D. T. Jones, and D. Roos (1991). *The Machine That Changed the World: The Story of Lean Production*, New York, Harper Perennial.

TACTICAL
CONSIDERATIONS

II

TACTICAL CONSIDERATIONS

Chapter 5

Designing the Reverse Logistics Network

Necati Aras, Tamer Boyacı, and Vedat Verter

Contents

5.1 Introduction

The reverse logistics (RL) network collects used products from end users; consolidates, inspects, and sorts them as needed; and transports them for various recovery options. It is therefore one of the most crucial components of closed-loop supply chains, from both environmental and financial perspectives. The quantity of collected products determines the amount of products, components, and raw material that can subsequently be remanufactured, reused, or recycled. The remainder enters the waste stream and ends up in increasingly capacitated landfill. On the one hand, there are significant costs associated with setting up and operating the logistical infrastructure for closed-loop supply chains. On the other hand, the recovery operations represent potential revenues or cost savings for firms. Chapter 9 provides some examples of existing profitable practices. Clearly, the profitability of these practices hinges on the effectiveness and efficiency of the underlying RL network.

In response to growing environmental concerns of the public and the resulting pressure from green organizations, governments around the globe have started to enact directives and pass legislation to reduce environmental damage caused by used products. As reviewed in Chapter 3, some of these regulations impose mandatory collection rates for end-of-life products, along with recovery and recycling targets. For example, in the context of electrical and electronic waste (WEEE), EU Directive (Directive 2003/108/EC) mandates a collection rate of at least 4 kg of WEEE per inhabitant per year, and depending on the product category, reuse, and recycling targets ranging from 50 to 75 percent by weight. Similarly, for end-of-life vehicles (ELV) Directive 2000/53/EC of the European Commission dictates the collection of all ELVs and mandates minimum recycling and recovery rates. Most legislation hold producers (manufacturer/importer) responsible for the costs of collection as well as the treatment, recovery, and disposal of their own products. Whether it is due to such legislation, social responsibility concerns, or potential economic benefits, more firms are adopting proactive approaches in closing the loop in their supply chains. Consequently, the design of the associated RL network is becoming increasingly important.

There are different RL network structures observed in practice. The nature of the used product and type of recovery has a major bearing on the structure. When used products have relatively high economic value and can be refurbished or remanufactured, original equipment manufacturers (OEMs) actively engage in used-product acquisition and recovery operations. This is commonly observed in the electronics industry. For example, IBM's Global Asset Recovery Services operates a wholly-owned global network of collection and refurbishment centers for recovering end-of-lease assets (e.g., servers, hard drives) and contracts with external recyclers for material recovery. Companies like HP and Xerox have similar initiatives. In this case, independent remanufacturers and refurbishers also actively pursue collection and recovery opportunities. When used products have relatively low economic value, which is more commonly the case for end-of-life products, take-back and recovery is often mandated by legislation. As highlighted in Chapter 3, take-back schemes are organized differently across the globe. In some countries, OEMs and importers have to deal with a nationwide non-profit organization that deals directly with recycling and treatment firms, calculates and charges the related costs to each OEM/importer. In the context of WEEE, such a system is used in countries including Belgium, the Netherlands, and Sweden. In contrast, in some other countries, OEMs/importers are free to establish their own network with recycling and treatment firms. European Recycling Platform, formed by Braun, Electrolux, HP, and Sony, offers WEEE compliance in ten EU member states including Austria, France, Germany, among others. A similar system is also operative in Japan. In either case, local authorities and municipalities contribute to the collection of end-of-life products by setting up public collection facilities. As the preceding discussion highlights, the RL network may involve multiple stakeholders including OEMs/importers (or a consortium of them) and possibly their forward distribution partners, third-party remanufacturers, recycling and treatment firms, third-party logistics firms, as well as local authorities.

This chapter identifies and discusses key strategic as well as operational issues involved with the design of the RL systems, provides an overview of existing approaches and results, with a special emphasis on business and managerial implications. To this end, we assume that the business case for recovery (remanufacturing, reuse, or recycling) has already been made. Chapter 2 of this book provides a coverage of strategic business concerns (e.g., presence of competition, technology choice) producers might have as they engage in recovery operations. We also adopt a centralized approach, where a single decision-maker is responsible for the design of the RL network. This enables us to take a comprehensive look at the underlying economics of designing logistics networks for reverse-supply chain operations. Although we are primarily concerned with the reverse supply chain, wherever appropriate, we emphasize the importance of linking the RL network with the forward distribution network.

We start the chapter by discussing some high-level decisions that need to be made with respect to the design of the RL network. We address questions including

- What is the right reverse channel structure? For example, should a producer use its existing retail network or a third-party firm to collect used products, or should it collect directly from end users themselves?
- What is the right collection strategy? Should the used products be picked-up from the end users or is it better to set up drop-off facilities for returns?
- Should financial incentives be given to entice the return of used products?
- How do financial incentives and the choice of the collection strategy influence the structure of the RL network?

We then proceed with the detailed design of the RL network. At this phase, the decision maker has knowledge on how the collection and recovery responsibilities are assigned, as well as the collection strategy to be used. By reviewing the relevant literature, we aim at shedding light on the following issues:

- Where to locate the facilities involved in an RL network, such as collection centers (CCs), inspection centers (ICs), remanufacturing facilities (RmF), and recycling facilities (RcF)?
- What are the flow patterns to be followed by the returned products through the RL network?
- What are the relevant tactical decisions such as acquisition prices and inventory levels, taken into account during the detailed design of the RL network?
- What is the impact of the uncertainties in market and operating conditions, such as demand and return levels?
- Should the reverse network be designed independently or jointly with the forward distribution network?

We provide an overview of existing academic literature in both areas, and discuss the related research results as well as their managerial implications. We remark that the methodologies employed for studying strategic design issues and the detailed design of the network typically differ. The former calls for simplified models that capture the essence of the high-level issues studied in a stylized and tractable manner, and are aimed at generating broad insights. In contrast, for the detailed design decisions, it is possible to develop more detailed and flexible models that can be adapted to different real-life scenarios as decision-support systems. In our discussion, we also highlight the variety of methodologies used in addressing the research questions mentioned above. Our chapter ends with a coverage of recent trends in RL network design, and an outlook on future research directions.

5.2 Strategic Design Issues

5.2.1 Reverse Channel Choice

Perhaps one of the first issues that arise in the design of an RL network is the decision as to who should take on the collection activity. This is a valid question even if the producer is ultimately financially responsible for collection and recovery.

There are various channel formats observed in practice. The forward distribution partners, given their proximity to the end market, are usually considered to be the ideal points for acquiring used products from end users. The classic example is Eastman Kodak, which receives single-use cameras from large retailers that also sell and develop films. Similarly, HP uses an authorized retail network to collect print cartridges (https://h30248.www3.hp.com/recycle/supplies/). In recent years, an increasing number of retailers have started their own collection initiatives to address growing concerns about the environment. For example, in the electronics sector, Best Buy, through a partnership with Greentec, accepts batteries, ink cartridges, CDs, and a number of portable electronics such as cell phones and MP3 players at its recycling points in the stores (http://www.bestbuy.ca/marketing/recycling/EN/). Similarly, in Japan end-of-use electronics products are collected via retailers (Dempsey et al. 2008).

In certain cases, producers prefer to collect directly from end users. For computer hardware, HP operates a program (https://warp1.external.hp.com/recycle/) that offers consumers and business owners the possibility to trade-in or receive cash refund for remanufacturable products and return older equipment for free collection. Similarly, Xerox collects end-of-lease copiers directly from customers as they install new ones (Savaskan et al. 2004). In other industries, collection is conducted by independent third parties. In the auto industry, for example, third-party dismantlers accept ELV and subsequently channel them for recycling and treatment.

As the preceding discussion highlights, there are various channel structures for reverse supply chain operations. Considering that there are significant costs associated with these operations as well as revenue or cost saving opportunities (at least for high-value returns that can be remanufactured), the channel structure can also have an effect on the forward supply chain prices. This has been initially noted in Savaskan et al. (2004). They consider a stylized model of a supply chain in which a producer sells new and remanufactured products through an independent retail channel. Remanufactured products are perfect substitutes of new products (e.g., single-use cameras), so any collected used product presents an opportunity to reduce average manufacturing costs. The collecting agent has to invest in advertising and promotions to induce a collection rate from customers, and there is diminishing return to the investment effort. There is also a variable cost of collection and handling returns, which is constant. Together, these imply a total collection cost structure that displays economies of scale (i.e., average cost of collection per unit decreases with the quantity of collected products). The demand in the product market is modeled as a downward sloping linear deterministic function of prices. In this decentralized setting, Savaskan et al. (2004) investigate three alternative reverse channel formats: (1) producer collects directly from the end customers, (2) producer contracts the collection to the retailer, and (3) producer contracts the collection to a third party. They characterize and compare the wholesale price, retail price, and collection rate under each format. Their analysis reveals that retail collection is optimal from the viewpoints of the producer, retailer, as well as the customers. Producers and retailers earn more profit, the product prices are lower and collection rates are higher under this channel structure. The intuition is that it is harder for the producer to coordinate prices and used-product return rates as it faces double marginalization (i.e., the price of the product includes the margins of both manufacturer and retailer) in the forward channel. By being closer to the final demand, the retailer can reflect the remanufacturing cost savings to the final product more efficiently.

The initial model in Savaskan et al. (2004) has been revisited recently by Atasu et al. (2009). In particular, they introduce a total collection cost structure also used by Ferguson and Toktay (2006) that displays diseconomies of scale, that is, average cost of collection per unit increases with the quantity of collected products, as it becomes costlier to increase collection rates. Under this form, they compare direct collection with retailer collection and find that the findings in Savaskan et al. (2004) are reversed. Atasu et al. (2009) argue that with diseconomies of scale, despite being closer to the market, the retailer does not efficiently reflect the cost savings from remanufacturing in the product price, and collects less than what the manufacturer would. They further show that the result holds even if new and remanufactured products are differentiated and are valued differently by the customer. Their results signify the importance of identifying the economies of scale factor in collection costs, as it has a critical impact on the optimal reverse supply chain structure.

The model in Savaskan et al. (2004) has also been extended to the case with competition in the retail market. Using a game-theoretic framework, Savaskan and Van Wassenhove (2006) compare direct collection with retail collection when the retailers sell substitutable products. They show that under direct collection, remanufacturing cost savings is the driver for the improvement in producer and supply chain profits. When the retailers are responsible for collection, the competition at the retail level intensifies, which can lead to lower retail prices, higher demand, and higher producer and supply chain profits. They show that when product substitutability is low, collection via retailers is preferred by the producers. On the other hand, when price competition is intense (high substitutability), direct collection is preferable.

A central assumption of these models is that the acquired used products result in cost savings for the producer. Although this may be the case for value-added recovery involving remanufacturing/refurbishing, products destined for material recycling may not lead to such savings. For such products, the main concern is reducing costs (Guide and Van Wassenhove 2001). There may also be multiple channels available for collecting used products. For example, in the context of WEEE, in addition to the three options (producer, retailer, third-party collection), nongovernmental organizations (NGOs), community organizations as well as municipal authorities might be involved in collection.

Motivated by the practices in the auto industry, Karakayalı et al. (2007) study the reverse channel choice for collecting and processing end-of-life durable products. The decentralized setting involves a collector who acquires the used products from the market, and a remanufacturer who recovers a part of the product and sells them in the service parts market and sends the remainder for material recycling. The supply of used products depends on the acquisition prices, while the demand for remanufactured parts depends on the selling price. Karakayalı et al. (2007) compare two reverse channel structures: (1) remanufacturer-driven channel where the producer outsources the remanufacturer activity and (2) collector-driven channel where the producer outsources the collection activity. They consider cases where the used products are of uniform quality as well as the case where there is heterogeneity in the quality of used products (and hence acquisition and selling prices).

They show that when the size of the used-product supply market is relatively larger (respectively, smaller) than the size of the remanufactured parts market, the producer prefers a collector-driven market (respectively, remanufacturer-driven market). They also identify the amount of investment the producer has to make (e.g., to improve salvage values) to meet the collection rates mandated by environmental legislation when the collection rates attained by the preferred channel falls short of these targets. Furthermore, a two-part tariff is proposed to coordinate the reverse channel and attain centralized profits.

We remark that earlier literature in this area assumes that the infrastructure for collecting used products already exists. Hence, fixed installation costs associated with setting up the collection network is ignored. Likewise, the transportation and logistics costs associated with moving collected products for inspection, sorting, and recovery are excluded. The inclusion of these costs (which may display different scale economies) is likely to have an impact on the reverse channel choice. Some of these costs are explicitly modeled in the following sections.

5.2.2 Collection Strategy Choice

There are two prevailing collection strategies observed in practice (McMillen 2001). Under a pick-up strategy, the products are collected from the end users, whereas under a drop-off strategy, the end users make the travel effort to a central point to return the product. A critical element of the RL network design is the delineation of the collection strategy to be used. This entails a careful assessment of the costs associated with each collection strategy, including both the fixed costs associated with facilities and the variable costs associated with logistics and transport activities. Continuous approximation is a powerful methodology to estimate these costs in a tractable manner, and generate broad economic insights on the preferability of each strategy under different design options.

The basic premise of continuous approximation methodology is to represent demand in a market area with a continuous function. In the context of reverse supply chains, the demand refers to products that are available for return. The underlying assumption is that the used products are not concentrated in specific (few) locations, but can be represented as a density that is assumed to be constant (or slowly varying) over the market area. Based on this assumption, it is possible to derive approximate unit costs per returned product. This approach has been originally developed for designing forward distribution systems (see Daganzo 1999 for a comprehensive review). Subsequently it has been used in the analysis of vehicle routing issues in RL systems (see Beullens et al. 2004 and the references therein). In what follows, we first provide a brief description of this approach and then discuss its application in our context.

Consider a market with a constant density of used products (denoted as φ) that are collected via pick-up strategy. Given a collection rate τ, this suggests the density of returned (collected) products is $\rho = \tau\varphi$. The annualized fixed cost of operating each CC is denoted as *F*. Suppose that each center serves a circular area with radius *r*. Then, the total number of collected products in that circle is the product of the density of collected products and the circle's area, or $\pi r^2 \rho$. This implies that the fixed cost per unit of collected product can be stated simply as

$$\frac{F}{\pi r^2 \rho}.$$

The logistics costs can be approximated by dividing the collection area into ring-radial zones with nearly rectangular pick-up areas, within which a single vehicle route originating and ending at the CC is optimized (Daganzo and Newell 1986). A sketch of this approach is depicted in Figure 5.1. There are two components of the logistics costs associated with these pick-up tours. The first component is the line-haul cost, which is the transportation cost of moving trucks from the collection facility to the start of a pick-up tour and from the end of the tour back to the collection facility. For a circular collection area with uniform product density, the average line-haul distance of a randomly selected tour can be computed as $2/3r$. Letting *c* denote the transportation cost per vehicle per distance and *v* denote the capacity of the trucks, assuming full-truck load tours, the average round-trip line-haul cost per product can be approximated as

$$\frac{4}{3}\frac{c}{v}r.$$

The second component is the vehicle routing cost associated with the pick-up tours. It is well known (Daganzo and Newell 1986) that under the square grid metric, the vehicle routing cost per product can be approximated as

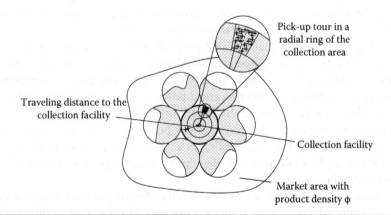

Figure 5.1 Continuous model for network design.

$$c\sqrt{\frac{1}{3\rho}}.$$

Hence the total fixed and variable costs per area under the pick-up strategy can be approximated as

$$\left(\frac{F}{\pi r^2 \rho} + \frac{4}{3}\frac{c}{v}r + c\sqrt{\frac{1}{3\rho}}\right)\rho.$$

Minimizing this cost over the radius r, the optimal size of the collection area can also be determined. From this it is possible to infer the approximate number of collection facilities to be located in the market area.

This network structure is referred to as local design in Fleischmann et al. (2004) and has been studied using the continuous approximation approach (see also Fleischmann 2003). In addition to these costs, they incorporate out-bound costs from the CC to an outside recovery facility (which could be done with larger capacity trucks) as well as disposal costs. They compare this structure with a central design where the products are collected and transported directly to a centralized recovery facility outside the market area. Such a design eliminates the need for CCs in the market area, but increases the transportation costs. Comparing the two design options under pick-up strategy, they determine a critical threshold distance around the recovery facility within which it is better to use a central design, whereas above this threshold a local design is preferable. The critical distance is increasing in the fixed cost and truck capacities, and decreasing in unit transportation cost and return density. Hence, high fixed cost structures and larger truck capacities favor a centralized design, whereas higher transportation costs and quantities of used products favor a local design. Fleischmann et al. (2004) also demonstrate the validity of continuous approximation method by comparing the average costs with those coming from a more detailed, discrete model.

The same continuous modeling approach can be used to estimate the costs under the drop-off strategy. Under the drop-off strategy, there would not be any logistics costs incurred within the service area as end users make the travel effort to the collection facilities, so the total cost (per unit collected) is composed of the annualized fixed cost and the out-bound transportation cost to the recovery facility (if any). Clearly, when the unit cost structure is the same for pick-up and drop-off strategies, under a constant rate of return, drop-off strategy would be the less-expensive strategy. Otherwise, the strategic choice between pick-up and drop-off options boils down to a simple comparison of the unit cost structures. This result, however, is heavily dependent on the constant return rate assumption, which implies that the amount of collected products does not depend on the accessibility of the collection network. Arguably, in reality, the further away the end user is from the nearest CC, the less willing he or she will be in dropping off the used products. Consequently, larger collection areas (i.e., less number of facilities within the market area) can result in lower collection rates. Accordingly, the return rate ρ is not constant, but depends on the service area (equivalently the radius r). In addition, the hassle of dropping off used-products can be alleviated by providing financial incentives, which can improve the return rates. Hence, the comparison between pick-up and drop-off strategies calls for a more detailed analysis that takes these issues into account. This is the subject of the next section.

5.2.3 Financial Incentives and Reverse Logistics Network Design

Price mechanisms can play a crucial role in acquiring used products from the market. This is especially important for high-value returns that can be remanufactured. Chapter 6 provides a detailed account of product acquisition management and specific pricing mechanisms. Our interest here is restricted to illustrating the general connection between financial incentives, collection strategy, and the design of the RL network.

Boyacı et al. (2008) applies the continuous modeling approach to the design of a collection network under a local design. They extend Fleischmann et al. (2004) by introducing the option of a drop-off strategy and allowing the product return rates to depend on the collection strategy in place, the accessibility of the network, as well as financial subsidies offered. Specifically, they assume that the collector offers a fixed subsidy s for returning a used product, while each return earns a constant revenue of p. These can be broadly interpreted as the average incentive and revenue per return respectively. They model the return decisions of end users using a utility-based choice model. An end user who is located at a distance x from the CC decides to return the used product when the associated utility $u_R(x,s)$ is above a reservation utility u_0 of not returning. The utility $u_R(x,s)$ is assumed to be linear increasing in the subsidy s and decreasing in travel distance x but is non-homogeneous across the population. Accordingly, the probability of return is calculated as $\Pr(x,s) = \Pr(u_R(x,s) > u_0)$. Integrating this over the circular collection area with radius r and product density φ, the average number of returns per area under drop-off strategy $\rho_{drop}(r,s)$ and pick-up strategy $\rho_{pick}(r,s)$ are obtained. Consequently, the expected profit per area under the pick-up strategy is approximated as

$$\Pi_{pick}(r,s) = \left(p - s - \frac{F}{\pi r^2 \rho_{pick}(r,s)} - \frac{4}{3}\frac{c}{v}r - c\sqrt{\frac{1}{3\rho_{pick}(r,s)}} \right)\rho_{pick}(r,s).$$

Similarly, under the drop-off strategy, the expected profit per area is estimated as

$$\Pi_{\text{drop}}(r,s) = \left(p - s - \frac{F}{\pi r^2 \rho_{\text{drop}}(r,s)} \right) \rho_{\text{drop}}(r,s).$$

Boyacı et al. (2008) analyze and compare the two functions in detail. They show that under the drop-off strategy, higher subsidies result in larger collection areas. This is because a higher subsidy increases the willingness to travel larger distances and hence the return rate, which justifies increasing the collection area (i.e., install-ing less facilities), and thereby save from the fixed costs. This implies that financial incentives and the number of collection facilities act as strategic substitutes in the acquisition of used products. In contrast, higher subsidies result in smaller col-lection areas (i.e., more collection facilities) under the pick-up strategy due to an increase in logistics costs. This implies that financial subsidy and the number of collection facilities are strategic complements with respect to the acquisition of used products. They also characterize the impact of fixed costs, logistics costs, used-product density in the market, as well as product characteristics such as bulkiness on the optimal subsidy and the collection network design under both strategies.

Comparing the profits under the pick-up and drop-off facilities, Boyacı et al. (2008) identify the drivers for the preference of each strategy. Regarding costs, they show the relative cost of obtaining a product is the key factor. For the pick-up strategy, the average cost of obtaining a product $\bar{c}_{\text{pick}}$ is determined mostly by the average line-haul cost per product. For the drop-off strategy, there are no direct costs, but the cost of obtaining a product $\bar{c}_{\text{drop}}$ can be estimated as the amount of subsidy that needs to be offered to induce an end user to travel a unit distance more. They find that the ratio $\bar{c}_{\text{pick}}/\bar{c}_{\text{drop}}$ governs the performance of each strategy. Furthermore, the dominance increases with higher fixed installation costs. They identify the used-product density as another driver. In particular, a high density of used-products (e.g., more urban areas) favors the use of a pick-up strategy, whereas drop-off strategy is more profitable for lower product densities (e.g., rural areas). Interestingly, neither the environmental awareness of the market (i.e., the overall willingness to participate in collection initiatives) nor the return value of the prod-uct has a major bearing on the collection strategy choice.

We remark that the literature discussed above is mainly concerned with the design of the reverse network. Fleischmann (2003) and Fleischmann et al. (2004) indicate that the forward distribution network can also be brought into the pic-ture using the continuous modeling approach. An explicit model is developed and analyzed by Wojanowski et al. (2007). Specifically, they study the design of a drop-off collection facility network in conjunction with a forward retail distribu-tion network, under a deposit–refund system. In such systems, the customer pays a deposit in addition to the price of the product, and the deposit is refunded when the used product is returned. As such, consumers make choices to purchase and whether to return the used product or not, which depend on the product price as well as the deposit–refund amount. They consider the design of the reverse network given an existing retail network, and also the integrated design case where retail and collection facilities are co-located. They identify the return value of the product as the main determinant of the amount collected. If the return value is high, a voluntary deposit–refund arrangement set by the collector can achieve a high collection rate. This is not true, however, for products with low return value. They show that for such products it may not be sufficient to impose minimum deposit–refund requirements; additional accessibility-based requirements may be necessary. Wojanowski et al. (2007) also show that it is optimal for the firm to subsidize a portion of the deposit in setting the retail price. This implies that it is not optimal to add the deposit onto the retail price.

5.3 Detailed Design of the Reverse Logistics Network

In the preceding section, we focused on some strategic design considerations and related models for the establishment of the RL network, especially related to the collection phase. Although these models are quite useful in generating guiding principles, they do not result in readily implementable designs for the RL network. Clearly, a more detailed design of an RL network must go beyond collection, and also determine the number and locations of ICs, remanufacturing and RcF. In this section, we provide an overview of the models and approaches used for this detailed design of the RL network. Figure 5.2 illustrates a generic closed-loop supply chain where solid arcs represent the forward flow of materials and dashed arcs represent the reverse flow of returned products. The facilities that belong to the RL network are depicted by shaded nodes. The possibility of co-locating

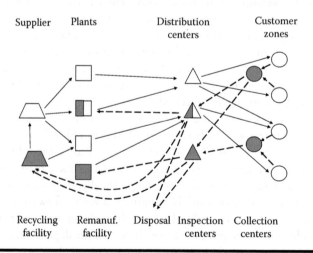

Supplier Plants Distribution centers Customer zones

Recycling facility Remanuf. facility Disposal Inspection centers Collection centers

Figure 5.2 A generic closed-loop supply chain.

forward and RL facilities is indicated by the half-shaded nodes. For example, the half-shaded square in the second echelon shows that an RmF is co-located with a manufacturing plant.

The majority of the academic literature on designing the RL network focuses on the location and configuration of the facilities that process returned products only. These RL network design papers mostly represent the flows from the customers toward upstream facilities although in some cases the forward flows of recovered products are also taken into account. In what follows, we first review studies that are concerned with only the "reverse network design" where no decisions are made regarding the structure of the forward supply chain. The papers that focus on establishing facilities in both the forward and reverse networks are subsequently reviewed under "integrated network design." A recent annotated bibliography of the literature on the reverse and integrated network design problems is provided by Akçalı et al. (2009).

5.3.1 Reverse Network Design

There has been considerable research on reverse network design. Starting with Spengler et al. (1997), we identified 21 refereed papers in this domain. These papers cover a wide scope of issues related to the design of the reverse network. The structural properties of the RL networks studied by these papers also vary significantly. For this reason, we do not find it helpful for the reader to provide a generic RL network design model that incorporates all the relevant aspects. Instead, we opt for a taxonomy of the existing models in terms of their major structural characteristics. These are

1. Depth of the RL network, that is, whether it contains CCs, ICs, RmF, and RcF
2. Tactical decisions incorporated
3. Existence of stochastic elements

Table 5.1 categorizes the RL network design papers in terms of these characteristics. We remark here that a consensus does not exist among the authors in terms of the terminology used for referring to RL facilities. For example, ICs that perform sorting and separation activities are also called return centers, intermediate centers, or disassembly centers, and RmF are sometimes named as treatment facilities or reprocessing facilities. In Table 5.1, we also indicate the solution approach/algorithm adopted and whether the proposed methodology is implemented in a real-life case.

We find the papers with case studies particularly relevant to practitioners. Therefore, we first focus on the RL network design articles that report on real-life applications in particular industries such as carpet, construction waste, steel by-products, and battery recycling, as well as the recovery activities for electronic and automotive industries. Then we review the papers with methodological contributions.

Table 5.1 Papers on Reverse Network Design

Paper	Depth of the Network				Tactical Decision	Stochasticity	Solution Algorithm	Case Study
	CC	IC	RmF	RcF				
Spengler et al. (1997)				X	Process selection	—	Exact, commercial solver	Constr. waste, steel by-products
Barros et al. (1998)		X		X	—	—	LP relaxation	Sand
Jayaraman et al. (1999)			X		—	—	Commercial solver	—
Krikke et al. (1999)			X	X	Inventory	—	Commercial solver	Copier
Louwers et al. (1999)				X	—	—	Exact	Carpet
Shih (2001)	X			X	—	—	Commercial solver	Appliances and PCs
Jayaraman et al. (2003)	X		X		—	—	Heuristic concentration	—
Schultmann et al. (2003)		X			—	—	Commercial solver	Battery
Realff et al. (2004)				X	Process selection	Return, price	Commercial solver	Carpet
Listeş and Dekker (2005)		X		X	—	Demand, return, cost	Exact	Sand
Min et al. (2006)	X	X			Inventory	—	Genetic algorithm	—

				Congestion level	Return, reman. time		
Lieckens and Vandaele (2007)			X	—	—	Genetic algorithm	—
Üster et al. (2007)	X		X	—	—	Exact	—
Aras and Aksen (2008)	X		X	Acquisition price	—	Tabu search	—
Aras et al. (2008)	X			Acquisition price, fleet size	—	Tabu search	—
Pati et al. (2008)	X	X		—	—	Commercial solver	Paper
Srivastava (2008)	X		X	—	—	Commercial solver	Appliances and PCs
Du and Evans (2008)			X	—	—	Scatter search	—
De Figueiredo and Mayerle (2008)	X			Acquisition price	—	Teitz and Bart heuristic	Tire
Cruz-Rivera and Ertel (2009)		X		—	—	Solver	Vehicle
Aksen et al. (2009)	X			Acquisition price, fleet size, subsidy	—	Tabu search	—

5.3.1.1 Papers with Case Studies

One of the bulky materials that occupy significant landfill space is disposed carpet. For example, in 1996, 1.6 million tons of carpet was disposed of annually in Western Europe (Louwers et al., 1999). This motivated the establishment of carpet recycling networks in Europe and North America. The current recycling technology allows for economic recovery of synthetic fibers from collected carpet waste. Louwers et al. (1999) present a planar location model to determine the best locations and capacities of regional carpet recovery centers incorporating reprocessing and transportation costs. Focusing on a U.S. application, Realff et al. (2004) make an explicit attempt to capture the uncertainty in synthetic fiber prices and return volumes. To this end, they first formulate a multi-period mixed-integer linear programming (MILP) model to determine the optimal sites for collection, reprocessing, and storage activities. Then, they incorporate the model in a robust optimization framework to minimize the maximum regret under nine plausible scenarios. Their analysis suggests that possible reductions in carpet collection volumes are a more significant threat for net revenues than possible reductions in fiber prices.

Another bulky material that needs to be redirected away from landfills is construction waste. The percentage of recycled construction waste has been increasing significantly over the years mainly due to legislative requirements. In their early work, Barros et al. (1998) concentrate on sieved sand that is a major by-product of recycled construction waste. In their case, sieved sand originates from 33 sorting facilities in the Netherlands. The regional depots receive the sieved sand and classify it as clean, half-clean, and polluted. The polluted sand is shipped to a treatment facility for cleaning and storage whereas regional centers store the clean and half-clean sand. The sieved sand recovery network works as a pull system where the demand of a number of construction sites around the country is served. Barros et al. (1998) develop a two-level capacitated facility location model to determine the number and locations of regional depots and treatment facilities. They observe that about half of the regional depots to be established are common under all scenarios considered, and these locations are close to the sources of sieved sand. Based on the same case, Listeş and Dekker (2005) formulate a stochastic programming (SP) formulation to incorporate uncertainties in the supply of sieved and in the demand of clean and half-clean sand. Assuming either a low supply or a high supply scenario, they develop a two-stage SP formulation where the locations of regional depots and treatment facilities are decided in the first stage, and one of the seven demand scenarios with equal probabilities is realized in the second stage. Considering the discrepancy between the high-supply and low-supply sieved sand volumes, Listeş and Dekker (2005) also develop a three-stage SP formulation where the location decisions are made during the first and second stages to also incorporate supply uncertainty.

Spengler et al. (1997) present a model for recycling the residues that are produced in large quantities during production of crude steel. The recycling of these by-products involves the reduction of the undesired materials such as zinc and lead. The authors aim at determining the optimal recycling process structure for each steel by-product as well as the locations and capacities of these processes to be installed.

More recently, Pati et al. (2008) consider paper recycling in India by examining a network consisting of five layers, that is, waste paper sources, dealers, godown owners, suppliers, and a manufacturer of recycled paper. The aim is to determine the most appropriate network partners at each level for the manufacturer with regard to three objectives: (1) RL cost, (2) separation of lower grade paper at the source, and (3) waste paper recovery. They use a priority goal programming formulation and investigate the impact of all possible priority rankings of the three objectives.

Used batteries constitute a threat for the environment due to potentially hazardous substances they contain. Therefore, there is an increasing effort around the globe for their collection and safe disposal. For example, Germany leads the EU by collecting over 10,000 tons annually, which amounts to a 30 percent collection rate. Although it translates into a lower volume, the collection rate of spent batteries in Belgium reaches 60 percent. The configuration of the German RL network for batteries is examined by Schultmann et al. (2003). The authors highlight the importance of the accuracy of the sorting process for batteries. In the event that the sorting process is inaccurate, the quality of the battery recycling process is jeopardized. They identify the number and locations of battery sorting facilities under two alternative scenarios. These scenarios differ from each other in terms of the percentage of mercury-free batteries available and the total weight to be collected.

Unrecoverable tires are among the most challenging streams of waste because of their volume and durability. According to the U.S. Environmental Protection Agency, it is estimated that on the average one tire per person reaches the end of useful life in North America annually. About 15 percent of discarded tires is reused for making retreaded tires for automobiles and trucks, whereas the remainder is recycled. An analytical model for designing a tire collection and recycling network in Southern Brazil is developed by De Figueiredo and Mayerle (2008). In their case, three million unrecoverable tires need to be collected by collecting agents from 682 municipalities, and shipped to a reprocessing facility through a set of receiving centers. The authors propose a bi-level mixed-integer nonlinear programming (MINLP) formulation from the perspective of a recycler who wishes to determine the optimal number and locations of receiving centers and the price to be paid to collecting agents per unrecoverable tire collected.

ELVs are among the most significant streams of returned products that consume landfill space unless properly disposed of. It is estimated that approximately 10–11 million ELVs in the United States and around 9 million ELVs in Europe arise annually. These vehicles need to be first dismantled to remove valuable parts for reconditioning, then shredded to recover ferrous and nonferrous metals for recycling. The recycling and recovery rates in 2000 were 75 percent by weight in the United States and Europe. The EU aims at increasing the recovery rate to 95 percent and the recycling rate to 85 percent by 2015 (Zoboli et al. 2000). Cruz-Rivera and Ertel (2009) study the RL network design for the collection of ELVs in Mexico. They use a simple plant location model for locating CCs under three different coverage scenarios.

Home appliances and computers account for a considerable portion of the returned products. In Europe, for example, 1.18 million tons of waste electrical and electronics equipment has been collected in 2007 (www.weee-forum.org). Large household appliances, air conditioners, TV sets, and computers comprise about 80 percent of this e-waste. Therefore, the establishment of RL systems to recover and recycle these materials has attracted the attention of the researchers and practitioners alike. Shih (2001) addresses the development of computer and home appliance collection and recycling network in Taiwan. The paper aims to determine the optimal sites for the storage and disassembly/recycling plants for the returned products. The reclaimed materials including copper, iron, and aluminum are sold at the material markets. Shih (2001) investigates six scenarios based on varying take-back rates and storage-sharing policies for the computers and home appliances, and identifies that the current number of storage sites exceeds the required number even under the high take-back rate. Recently, Srivastava (2008) discusses the RL network design issues pertaining to electronic products and appliances in India. The lack of access to the state-of-the-art remanufacturing technologies and large capital investments required for these technologies seem to be the main bottlenecks for widespread implementation of remanufacturing in India. In the European context, Krikke et al. (1999) analyze an RL network redesign initiative at Océ, a copier manufacturer in the Netherlands. Using a detailed MILP formulation, the authors compare three alternative RL network designs and find out that the cost differences are fairly small. They point out that the decision needs to be justified by the firm's business strategy.

5.3.1.2 Methodological Papers

In an early work, Jayaraman et al. (1999) develop an MILP model for a basic RL network where collected cores of different types are remanufactured and sent back to customer zones who demand remanufactured products. The model optimizes the number and locations of the RmF that also serve as storage sites. The proposed model is solved via a commercial solver using a number of illustrative problem instances.

Jayaraman et al. (2003) formulate an MILP model in which a given number of products returned to retail outlets are first sent to collection facilities and then transshipped to refurbishing facilities. The objective of their model is to find the optimal number and location of these two types of facilities. They develop a heuristic framework where heuristic concentration module for finding the most likely sites of collection and refurbishing facilities is complemented with a heuristic expansion procedure. They solve problems with up to 100 retail sites, 40 potential collection sites, and 30 potential refurbishing sites.

Another paper that deals with collection facilities explicitly is Min et al. (2006), which presents an MINLP model to determine the optimal number and locations of collection points as well as centralized return centers. The proposed model also optimizes the length of a collection period at each collection facility so as to incorporate inventory holding costs. The authors develop a heuristic based on genetic algorithms. Using hypothetical parameter values, they find that the total numbers of collection facilities and centralized return centers are robust, whereas the total logistics cost is sensitive to inventory-related decisions. In particular, the maximum collection period at the collection facilities and the inventory holding costs seem to have an important effect on the total cost.

Du and Evans (2008) focus on the design of an RL network for returned items that need repair. The flow of returned products from the collection sites to the repair facilities must be balanced with the flow of spare parts shipped from the manufacturing plants. Considering the importance of customer service in the context of repairs, the authors formulate a bi-objective MILP model to incorporate tardiness in the cycle time as well as the total cost. The decisions to be made consist of the locations and capacity levels of repair facilities. The constraint method is employed to convert the formulation to an MILP that is solved via the scatter search algorithm. A set of nondominated solutions is generated by iteratively tightening the upper bound on one of the objectives.

Aras and Aksen (2008) analyze an uncapacitated CC location problem (CCLP) for incentive- and distance-dependent returns. In their profit maximization model, a drop-off policy is in effect. Their decision whether or not to participate in this buyback campaign is affected by the distance to the nearest CC and the financial incentive that depends on the quality state of the used product. The authors propose two MINLP models for the fixed-charge and p-median versions of the CCLP. In a later paper, Aras et al. (2008) work on the p-median version of the same CCLP under a pickup policy in which all collection related costs, that is, the cost of operating the vehicles and transportation cost are incurred by the collecting company. The aim is to determine the locations of the CCs, the level of the financial incentive, as well as the number and load mix of the vehicles. In both papers, the authors utilize Tabu search–based solution procedures.

Most recently, Aksen et al. (2009) present a bi-level formulation framework describing the subsidy agreement between the government (the leader) and a company engaged in collection and recovery operations (the follower). The authors study two alternative policies. The first is a supportive policy where the government uses monetary incentives to motivate the achievement of a target collection rate by the company. The second is a legislative policy where the government mandates a certain collection target while ensuring economic viability of the company. In both cases, the government minimizes the required subsidy per collected item and the company maximizes its profit. They show that at the same profitability and collection levels, lower subsidy levels are required under the legislative policy.

Üster et al. (2007) apply Benders decomposition method for solving a multi-product RL network design problem. For a given set of new product plants and distribution centers (DCs), they determine the locations of CCs and the remanu-factured product plants so as to minimize the total forward and RL costs. They assume that each retailer works with a single DC and a single CC for all of its new and returned products, respectively. In addition, the new product plants are prod-uct dedicated and exactly one remanufactured product plant can be established for each type of returned product. These single-sourcing assumptions facilitate the use of Benders decomposition and enable the authors to generate alternative Benders cuts via different separations at the subproblem level. On the basis of experiments performed on hypothetical instances, they find that Benders cuts based on flow and product separation are the most effective. Also, they observe that the more chal-lenging instances are those where the contribution of different cost components to the overall objective value is balanced.

The only paper in this group that incorporates uncertainty is by Lieckens and Vandaele (2007). They embed queuing constructs in an RL network design model to capture the congestion in the RmF. The objective of the model is to find the location and capacity level of each RmF to be installed. The amount of returns col-lected and the reprocessing times are uncertain. The arising MINLP formulation is tackled by a genetic algorithm-based differential evolution technique.

5.3.2 Integrated Network Design

There has been relatively less interest on the integrated design of forward and reverse networks. We identify 14 papers in this domain. In addition to the four reverse facility types (CC, IC, RmF, and RcF), these papers also study the establishment of manufacturing plants (P) and DCs. Table 5.2 presents a categorization of these papers utilizing the same attributes as in the previous section. Here, we also start with an overview of the papers that contain case studies, and continue with a brief account of the studies that make methodological contributions.

5.3.2.1 Papers with Case Studies

One of the earliest integrated RL design models inspired by real-life applications is due to Fleischmann et al. (2001). In their model, the facilities are (1) plants where both manufacturing of brand new products and remanufacturing of used products are performed, (2) warehouses that act as transshipment points between plants and customer locations, and (3) disassembly centers that perform inspec-tion on returned products that are collected at the customer locations and then shipped to these centers. In addition to the transportation costs of goods and fixed costs of opening facilities, the objective function of the MILP model also

Table 5.2 Papers on Integrated Network Design

Paper	Depth of the Network							Tactical Decision	Stochasticity	Solution Algorithm	Case Study
	CC	IC	RmF	RcF	P	DC					
Marín and Pelegrin (1998)	X				X		—	—	Exact	—
Fleischmann et al. (2001)		X	X		X	X	—	—	Commercial solver	Paper, copier
Krikke et al. (2003)		X	X	X	X	X	—	—	Commercial solver	Refrigerator
Beamon and Fernandes (2004)		X				X	—	—	Exact	—
Salema et al. (2006)		X			X	X	—	—	Commercial solver	Office document
Salema et al. (2007)		X			X	X	—	Demand, return	Commercial solver	Office document
Ko and Evans (2007)		X				X	—	—	GA, commercial solver	—
Lu and Bostel (2007)		X	X		X		—	—	Lagrangean heuristic	—

(continued)

Table 5.2 (continued)   Papers on Integrated Network Design

Paper	Depth of the Network						Tactical Decision	Stochasticity	Solution Algorithm	Case Study
	CC	IC	RmF	RcF	P	DC				
Sahyouni et al. (2007)	X					X	—	—	Lagrangean heuristic	—
Listeş (2007)		X			X		—	Demand, return	Exact	—
Demirel and Gökçen (2008)	X	X				X	—	—	Commercial solver	—
Lee and Dong (2008)	X					X	—	—	Tabu search	—
Zhou and Wang (2008)	X		X		X	X	—	—	Commercial solver	—
Lee and Dong (2009)	X					X	—	Demand, return	Heuristic	—
Salema et al. (2009)		X			X	X	Inventory	—	Commercial solver	Office document

includes the cost of unsatisfied demand, the cost of uncollected used products, and the cost savings associated with remanufacturing. On the basis of this model, the authors compare the sequential and integrated approaches to design decisions. In the sequential design approach, the solution to the model with forward flows (i.e., the locations of plants and warehouses) is prespecified when deciding the reverse network structure (i.e., the locations of the disassembly centers). Note that this represents the decision process of a firm with an already established forward distribution channel. A general-purpose MILP solver, CPLEX, is used to solve the models in both approaches. Analyzing two examples inspired by real-life industrial cases, copier remanufacturing and paper recycling, it is concluded that the reverse flows have a significant impact on the overall network structure only when the forward and reverse channels differ in a considerable way with respect to geographical distribution of demand and supply sites or cost structure. Otherwise, the fixed forward network structure does not impose important restrictions on the design of the reverse network. The authors also point out that return volumes constitute a key factor in the design decisions.

The formulation in Fleischmann et al. (2001) uses path-based flow variables corresponding to each plant–warehouse–customer triplet. Salema et al. (2006) offer an alternative arc-based formulation where flow variables are defined for plant–warehouse and warehouse–customer pairs. They suggest that their formulation is more effective as it contains less continuous decision variables, and hence larger instances can be solved more efficiently via commercial software. An extended model is also provided for the capacitated and multiproduct version of the design problem, which is implemented for two products. Salema et al. (2006) highlight the effectiveness of their formulations using two case studies. The first case is based on a document-office company in Spain where the alternative facility sites and customer zones are provided by the company whereas the costs and demand/return volumes are hypothetical. The second case is copier remanufacturing in Europe originally studied by Fleischmann et al. (2001). The capacitated and multiproduct extensions of Fleischmann et al. (2001) are considered in Salema et al. (2007). Based on a case with two products and three scenarios, they employ a scenario-based approach to incorporate the impact of demand and return uncertainty on logistics network design. Salema et al. (2009) is an effort to incorporate tactical decisions, such as production and inventory levels, in the integrated RL network design. To this end, they use two different time scales where the network design decisions are made during the macro time periods and tactical decisions are made during the micro time periods. Extending the arc-based formulation of Salema et al. (2006) to represent a set of macro and micro time periods, Salema et al. (2009) demonstrate that the arising model can handle fairly large problem instances encountered in practice. Interestingly, their model prescribes a zero stock policy for the Iberian company case.

In perhaps the most detailed case study for integrated RL network design, Krikke et al. (2003) focus on the forward and reverse supply chains of refrigerators. They evaluate three alternative refrigerator designs from the perspective of their overall costs, energy consumption, and waste generated. A goal programming style formulation is presented to minimize the total weighted deviation from predetermined targets for the three objectives. The bill of material for a refrigerator is represented at three levels, that is, component, model, and product. The model includes manufacturing processes at each level, warehouses as well as facilities for repair, disassembly, inspection, rebuild, and recycling. An MILP formulation is presented where each of the activities mentioned above are assigned to alternative sites. Based on a detailed analysis of the refrigerator case, Krikke et al. (2003) make the following observations. The overall network design has a clear impact on costs whereas the product design is more influential on energy consumption and waste generated. It seems that modular refrigerator designs are effective means of making the trade-off among cost, energy use, and waste objectives. The reuse of components and modules comprises the most beneficial recovery option. In terms of cost, centralization is a better strategy than decentralization.

5.3.2.2 Methodological Papers

In an early effort, Marín and Pelegrin (1998) extend the simple plant location problem to define the return plant location problem where each manufacturing plant to be established also serves as a CC for customer returns. This is presumably the most basic integrated RL network design problem where the sets of potential sites for manufacturing plants and potential sites for CCs are the same. The authors assume that the number of returns is proportional to the demand of each customer and the remanufacturing capacity of a plant is proportional to its manufacturing capacity. They develop a heuristic solution procedure based on Lagrangean decomposition as well as an exact procedure based on branch-and-bound. A more detailed network is considered in Beamon and Fernandes (2004) where the manufacturing plants serve the customer demand via warehouses and receive the returns via CCs and warehouses. The aim is of the model is to determine the best locations of warehouses with and without sorting capability as well as CCs. Recently, Zhou and Wang (2008) present a very similar model to that of Fleischmann et al. (2001) in which returns can be repaired at centralized return centers and sent back to the warehouses to satisfy the customer demand. In a similar modeling framework, Demirel and Gökçen (2008) allow for the direct shipment of the returns from the customer zones to the disassembly centers bypassing the CCs.

Several authors developed heuristic approaches for solving a variety of integrated RL network design formulations. Below, we outline four such papers. Lu and Bostel (2007) provide an MILP formulation where the customer zones are directly served from manufacturing or RmF, whereas reverse flows go through intermediate centers, which perform cleaning, disassembly, testing, and sorting, to the RmF. Their path-based model is solved by Lagrangean relaxation of three sets of constraints stipulating that (1) customer demand must be satisfied, (2) all returns must be collected, and (3) an intermediate center can be used only if the associated fixed costs are incurred. The authors suggest the use of additional constraints to strengthen the relaxation. Based on problem instances derived from classical location problems, Lu and Bostel (2007) show that their solution procedure outperforms CPLEX in terms of solution accuracy and efficiency. Sahyouni et al. (2007) also formulate an MILP model where the customer demand is served from DCs and customer returns are collected at CCs. The model allows for establishing hybrid facilities that handle both forward and reverse flows. Sahyouni et al. (2007) constitutes an extension of Marín and Pelegrin (1998) where all facilities belong to both the forward and reverse networks. As the solutions approach, the authors employ Lagrangean relaxation. A practically relevant feature of this paper is the network similarity metric that can be used for comparing alternative network designs.

Ko and Evans (2007) addresses the problem of a third-party logistics (3PL) provider who runs the warehouses and repair centers performing inspection and separation activities. The client company operates a set of existing plants that aim at satisfying the market demand via the warehouses and collecting the returns through the repair centers. The forward and RL activities of the 3PL company can be colocated to achieve cost savings. Ko and Evans (2007) develop a multi-period model to determine the opening, expansion, and closing decisions of the warehouses and repair centers over time. The resulting MINLP model is solved by means of a genetic algorithm. Min and Ko (2008) present a very similar model to that of Ko and Evans (2007). Lee and Dong (2008) develop a Tabu search–based heuristic for integrated RL network design in end-of-lease computers. In their model, there is a single OEM who wants to establish a set of capacitated hybrid processing facilities that serve as both warehouses and CCs.

There are two papers that make an explicit attempt to incorporate the uncertainty in demand and return volumes. Listeş (2007) provides a scenario-based formulation for an RL network design problem in which plants and ICs are located and transport links are established. The objective is to minimize the total cost of establishing and operating the network less the expected revenue that depends on the uncertain demand and return volumes. Listeş (2007) implements the integer L-shaped method as an efficient decomposition approach for solving the resulting MILP formulation. Using an illustrative example consisting of 12 scenarios, he demonstrates that the stochastic network design can be different from any of the designs that are based on alternative scenarios. Lee and Dong (2009) extend their earlier work to a multi-period setting where the locations of forward, return, and hybrid processing facilities are determined. Based on their previous deterministic model, they develop a two-stage SP formulation with demand and return uncertainties. The location decisions are made at the first stage while the second stage optimizes the flow decisions based on the realization of the uncertain parameters. A simulated annealing–based heuristic algorithm is combined with a sample average approximation scheme to generate a solution procedure.

5.4 Conclusions and Outlook

As the preceding review highlights, the academic literature on the detailed design of RL network typically lacks deep managerial insights that could be of immediate help to practitioners. This can be attributed to industry specificity and the solution-focused approach taken by these papers, as well as the structural properties of the mathematical formulations utilized for detailed design. Needless to say, it is crucial that in addition to methodological contributions, the research in this field should address questions that are relevant in practice. For example, is there a significant difference between integrated and sequential design alternatives for integrated RL network design problems in terms of cost and solution structure? If so, what is the scale of the benefits associated with each design approach and what are the underlying drivers? Likewise, should inspection be carried out at separate centers or at the recovery facility? Under what conditions would the integration of inspection and recovery would be beneficial?

To this end, our ongoing work makes an explicit attempt to address these issues (Verter and Aras 2008). The modeling framework involves the determination of the optimal number and location of the DCs and ICs so as to minimize the total cost of establishing and operating the closed-loop network, given a set of existing manufacturing and remanufacturing plants and associated capacities. Under the integrated design option, DC/IC configuration and flows are simultaneously determined. This is compared against the sequential design option, which involves first making the DC location and forward flow decisions without incorporating the reverse flows, and then configuring the reverse supply chain by taking the forward chain structure as given. The computational results obtained on a large number of randomly generated instances indicate that the cost advantages of the integrated design can easily exceed 10 percent (the reported maximum is 14 percent). Interestingly, the reverse network structure seems to be robust to the design approach and the cost difference is mainly due to the forward network configuration. This suggests that the ability of the forward network to adapt itself to the presence of reverse flows can be the main advantage of the integrated design approach. In the event that the firm already has an established forward network, the integrated solution can serve as a target configuration for the existing DCs to converge in the long run.

Verter and Aras (2008) also identify the level of remanufacturing capacity utilization as a key determinant of the potential benefits that can be achieved via the integrated approach. This relates to the ability of the firm to ship the recoverable returns to the RmF at the plants with cheaper DC connections. Furthermore, the benefits of integrated design first increase and then decrease as the return ratio increases. For high values of the return ratio, the cost difference between the integrated and sequential approaches is minimal because the forward and reverse network structures become similar (as also observed in Fleischmann et al. 2001). Consequently, the integrated approach is most beneficial for medium values of the return ratio.

In the basic framework of Verter and Aras (2008), the ICs are located in the same echelon as the DCs. Alternatively, ICs can be colocated with RmF in the upper echelon. This would save fixed costs due to economies scale in colocation, but would incur additional transportation costs because the recoverable returns are no longer disposed of early in the reverse supply chain. Analyzing an extension that finds the optimal location of RmF and allows for colocation of ICs, it is shown that the benefit associated with the integration of inspection, separation, and recovery improves as the overall quality of returns increases.

There are practically relevant issues related to the strategic design of the RL network that deserve more research as well. For example, the existing literature considers the exclusive use of the pickup or the drop-off collection strategy in the entire RL network. As also argued in this chapter, each strategy has its advantages and disadvantages. Hence, a hybrid strategy involving both pickup and drop-off is likely to outperform both. In particular, it might be possible to set up drop-off facilities serving relatively small zones, making the RL network accessible and increasing the collection rate. These products can then be picked up and transported to consolidation or recovery facilities. Critical issues here would be the determination of the optimal drop-off/pickup boundary and the potential benefits of using a hybrid strategy. Existing research also signifies the link between financial incentives and the RL network. Depending also on the collection strategy in place, these incentives can complement or substitute the accessibility provided by the RL network, and hence have significant cost and profit implications. The current research in this area considers simple financial incentive mechanisms (e.g., fixed subsidy per return). There is a need to consider finer mechanisms that can differentiate the incentive based on the product, its condition, and possibly other factors. In effect, this calls for the bridging of product acquisition management with the design of the RL network.

As environmental sustainability is gaining more attention, new regulations are being planned or implemented across the globe. There is a parallel growing stream of works in operations management addressing the impact of such legislation (e.g., take-back legislation) on closed-loop supply chains. The RL network constitutes a vital component of the closed-loop supply chain. Prevalent literature recognizes the importance of environmental legislation, however, further research is necessary to understand the precise impact such legislation and different policy tools have on the structure of the RL network and the operating economics. Ideally, this research should provide critical insights to policy makers and help the shaping of future legislation, balancing environmental objectives with economic and social ones.

On a related matter, it is evident from the papers reviewed in this chapter that the significant majority of the research on RL network design considers economic objectives (cost minimization or profit maximization), without paying too much attention to environmental impact. Environmental legislation and the pressure from consumers and NGOs are making it imperative for firms to mitigate the environmental impact of their operations. With emissions trading becoming more widely implemented (e.g., EU Emission Trading System for CO_2 emissions), there will also be direct costs associated with environmental performance. Accordingly, there is a need to integrate environmental performance in the design of the forward and RL networks. Unfortunately, there is no single measure that captures environmental impact in a comprehensive manner. Carbon emissions, cumulative energy demand, amount of toxic material released, effects on ozone layer depletion or global warming, among others, constitute alternative measures for assessing environmental impact. Hence, it is necessary to develop appropriate, quantifiable metrics for measuring environmental performance. Further research is necessary in developing integrated frameworks for design that take into account these multiple factors and multiple objectives. New constraints such as caps on the CO_2 emissions may need to be introduced. By surfacing the trade-offs associated with economic and environmental performance, these integrated frameworks would provide invaluable decision support to practitioners, and assist them in balancing the two objectives. Some research in this area has already started. For example, using the European pulp and paper industry as the background, Quariguasi Frota Neto et al. (2008) present a multi-objective formulation for designing the logistics network considering cost and environmental impact. They determine the efficient frontier, and show that the current state has room for improving both objectives. We anticipate a significant amount of research activity in this domain in the near future.

References

Akçalı, E., S. Cetinkaya, and H. Üster. 2009. Network design for reverse and closed-loop supply chains: An annotated bibliography of models and solution approaches. *Networks* 53(3), 231–248.

Aksen, D., N. Aras, and A. G. Karaarslan. 2009. Design and analysis of government subsidized collection systems for incentive-dependent returns. *International Journal of Production Economics* 119(2), 308–327.

Aras, N. and D. Aksen. 2008. Locating collection centers for distance- and incentive-dependent returns. *International Journal of Production Economics* 111(2), 316–333.

Aras, N., D. Aksen, and A. G. Tanuğur. 2008. Locating collection centers for incentive-dependent returns under a pick-up policy with capacitated vehicles. *European Journal of Operational Research* 191(3), 1223–1240.

Atasu, A., B. Toktay, and L. N. Van Wassenhove. 2009. The Impact of Collection Cost Structure on Reverse Channel Choice. *Working Paper*. College of Management, Georgia Institute of Technology, Atlanta, GA.

Barros, A. I., R. Dekker, and V. Scholten. 1998. A two level network for recycling sand: A case study. *European Journal of Operational Research* 110(2), 199–214.

Beamon, B. M. and C. Fernandes. 2004. Supply-chain network configuration for product recovery. *Production Planning and Control* 15(3), 270–281.

Beullens, P., D. Van Oudheusden, and L. N. Van Wassenhove. 2004. Collection and vehicle routing issues in reverse logistics. In *Reverse Logistics: Quantitative Models for Closed-Loop Supply Chains*, R. Dekker, M. Fleischmann, K. Inderfurth, and L. N. Van Wassenhove (Eds.). Springer-Verlag, Berlin, Germany.

Boyacı, T., V. Verter, F. Toyasaki, and R. Wojanowski. 2008. Collection System Design, Strategy Choice and Financial Incentives for Product Recovery. *Working Paper.* Desautels Faculty of Management, McGill University, Montreal, Canada.

Cruz-Rivera R. and J. Ertel. 2009. Reverse logistics network design for the collection of end-of-life vehicles in Mexico. *European Journal of Operational Research* 196(3), 930–939.

Daganzo, C. F. 1999. *Logistics Systems Analysis*, 3rd edn. Springer-Verlag, Heidelberg, Germany.

Daganzo, C. F. and G. F. Newell. 1986. Design of multiple-vehicle delivery tours-I: A ring-radial network. *Transportation Research-B* 20B(5), 345–363.

De Figueiredo, J. N. and S. F. Mayerle. 2008. Designing minimum-cost recycling collection networks with required throughput. *Transportation Research Part E* 44(5), 731–752.

Demirel, Ö. N. and H. Gökçen. 2008. A mixed-integer programming model for remanufacturing in reverse logistics environment. *International Journal of Advanced Manufacturing Technology* 39(11–12), 1197–1206.

Dempsey, M., C. Van Rossem, R. Lifset, J. Linnell, J. Gregory, A. Atasu, J. Perry et al. 2008. Developing Practical Approaches for Individual Producer Responsibility: Industry Report to the European Commission.

Du, F. and G. W. Evans. 2008. A bi-objective reverse logistics network analysis for post-sale service. *Computers and Operations Research* 35(8), 2617–2634.

Ferguson, M. E. and B. Toktay. 2006. The effect of competition on recovery strategies. *Production and Operations Management* 15(3), 351–368.

Fleischmann, M. 2003. Reverse logistics network structures and design. In *Business Aspects of Closed-Loop Supply Chains*, V. D. R. Guide Jr. and L. N. Van Wassenhove (Eds.). Carnegie Mellon University Press, Pittsburgh, PA, pp. 117–148.

Fleischmann, M., P. Beullens, J. M. Bloemhof-Ruwaard, and L. N. Van Wassenhove. 2001. The impact of product recovery on logistics network design. *Production and Operations Management* 10(2), 156–173.

Fleischmann, M., J. M. Bloemhof-Ruwaard, P. Beullens, and R. Dekker. 2004. Reverse logistics network design. In *Reverse Logistics: Quantitative Models for Closed-Loop Supply Chains*, R. Dekker, M. Fleischmann, K. Inderfurth, and L. N. Van Wassenhove (Eds.). Springer-Verlag, Berlin, Germany.

Guide Jr., V. D. R. and L. N. Van Wassenhove. 2001. Managing product returns for remanufacturing. *Production and Operations Management* 10(2), 142–155.

Jayaraman, V., V. D. R. Guide, and R. Srivastava. 1999. A closed-loop logistics model for remanufacturing. *Journal of the Operational Research Society* 50(5), 497–508.

Jayaraman, V., R. A. Patterson, and E. Rolland. 2003. The design of reverse distribution networks: Models and solution procedures. *European Journal of Operational Research* 150(1), 128–149.

Karakayalı, İ., H. Emir-Farinas, and E. Akçalı. 2007. An analysis of decentralized collection and processing of end-of-life products. *Journal of Operations Management* 25(6), 1161–1183.

Ko, H. J. and G. W. Evans. 2007. A genetic-based heuristic for the dynamic integrated forward/reverse logistics network for 3PLs. *Computers and Operations Research* 34(2), 346–366.

Krikke, H. R., A. van Harten, and P. C. Schur. 1999. Business case Océ: Reverse logistic network redesign for copiers. *OR Spectrum* 21(3), 381–409.

Krikke, H. R., J. M. Bloemhof-Ruwaard, and L. N. Van Wassenhove. 2003. Concurrent product and closed-loop supply chain design with an application to refrigerators. *International Journal of Production Research* 41(16), 3689–3719.

Lee, D.-H. and M. Dong. 2008. A heuristic approach to logistics network design for end-of-lease computer products recovery. *Transportation Research Part E* 44(3), 455–474.

Lee, D.-H. and M. Dong. 2009. Dynamic network design for reverse logistics operations under uncertainty. *Transportation Research Part E* 45(1), 61–71.

Lieckens, K. and N. Vandaele. 2007. Reverse logistics network design with stochastic lead times. *Computers and Operations Research* 34(2), 395–416.

Listeş, O. 2007. A generic stochastic model for supply-and-return network design. *Computers and Operations Research* 34(2), 417–442.

Listeş, O. and R. Dekker. 2005. A stochastic approach to a case study for product recovery network design. *European Journal of Operational Research* 160(1), 268–287.

Louwers, D., B. J. Kip, E. Peters, F. Souren, and S. D. P. Flapper. 1999. A facility location allocation model for reusing carpet materials. *Computers and Industrial Engineering* 36(4), 855–869.

Lu, Z. and N. Bostel. 2007. A facility location model for logistics systems including reverse flows: The case of remanufacturing activities. *Computers and Operations Research* 34(2), 299–323.

Marín, A. and B. Pelegrin. 1998. The return plant location problem: Modelling and resolution. *European Journal of Operational Research* 104(2), 375–392.

McMillen, A. 2001. Separation and collection systems. In *The McGraw-Hill Recycling Handbook*, H. Lund, (Ed.). McGraw-Hill Publishers, New York.

Min, H. and H. J. Ko. 2008. The dynamic design of a reverse logistics network from the perspective of third-party logistics service providers. *International Journal of Production Economics* 113(1), 176–192.

Min, H., H. J. Ko, and C. S. Ko. 2006. A genetic algorithm approach to developing the multi-echelon reverse logistics network for product returns. *Omega* 34(1), 56–69.

Pati, R. K., P. Vrat, and P. Kumar. 2008. A goal programming model for paper recycling system. *Omega* 36(3), 405–417.

Quariguasi Frota Neto, J., J. M. Bloemhof-Ruwaard, J. A. E. E. van Nunen, and E. van Heck. 2008. Designing and evaluating sustainable logistics networks. *International Journal of Production Economics* 111(2), 195–208.

Realff, M. J., J. C. Ammons, and D. Newton. 2004. Robust reverse production system design for carpet recycling. *IIE Transactions* 36(8), 767–776.

Sahyouni, K., C. Savaskan, and M. Daskin. 2007. A facility location model for bidirectional flows. *Transportation Science* 41(4), 484–499.

Salema, M. I., A. P. B. Póvoa, and A. Q. Novais. 2006. A warehouse-based design model for reverse logistics. *Journal of the Operational Research Society* 57(6), 615–629.

Salema, M. I., A. P. B. Póvoa, and A. Q. Novais. 2007. An optimization model for the design of a capacitated multi-product reverse logistics network with uncertainty. *European Journal of Operational Research* 179(3), 1063–1077.

Salema, M. I., A. P. B. Póvoa, and A. Q. Novais. 2009. A strategic and tactical model for closed-loop supply chains. *OR Spectrum* 31(3), 573–599.

Savaskan, C. and L. N. Van Wassenhove. 2006. Reverse channel design: The case of competing retailers. *Management Science* 52(1), 239–252.

Savaskan, C., S. Bhattacharya, and L. N. Van Wassenhove. 2004. Closed-loop supply chain models with product remanufacturing. *Management Science* 50(2), 239–252.

Schultmann, F., B. Engels, and O. Rentz. 2003. Closed-loop supply chains for spent batteries. *Interfaces* 33(6), 57–71.

Shih, L. S. 2001. Reverse logistics system planning for recycling electrical appliances and computers in Taiwan. *Resources, Conservation and Recycling* 32(1), 55–72.

Spengler, T., H. Püchert, T. Penkuhn, and O. Rentz. 1997. Environmental integrated production and recycling management. *European Journal of Operational Research* 97(2), 308–326.

Srivastava, S. K. 2008. Network design for reverse logistics. *Omega* 36(4), 535–548.

Üster, H., G. Easwaran, E. Akçalı, and S. Çetinkaya. 2007. Benders decomposition with alternative multiple cuts for a multi-product closed-loop supply chain network design model. *Naval Research Logistics* 54(8), 890–907.

Verter, V. and N. Aras. 2008. Designing Distribution Systems with Reverse Flows. *Working Paper*. Desautels Faculty of Management, McGill University, Montreal, Canada.

Wojanowski, R., V. Verter, and T. Boyacı. 2007. Retail collection network design under deposit refund. *Computers and Operations Research* 34(2), 324–345.

Zhou, Y. and S. Wang. 2008. Generic model of reverse logistics network design. *Journal of Transportation Systems Engineering and Information Technology* 8(3), 71–78.

Zoboli, R., G. Barbiroli, R. Leoncini, M. Mazzanti, and S. Montresor. 2000. *Regulation and Innovation in the Area of End-of-Life Vehicles*, F. Leone and DG JRC-IPTS (Eds.). IDSE-CNR, Milan, Italy.

Chapter 6

Product Acquisition, Grading, and Disposition Decisions

Moritz Fleischmann, Michael R. Galbreth,
and George Tagaras

Contents

6.1 Introduction

As for any supply chain, one of the main tasks of a closed-loop supply chain (CLSC) is to match supply with demand. Being able to supply goods at a lower cost than what the customer is willing to pay is what drives the economic viability of a supply chain. In a CLSC, this task raises particular issues on the supply side, due to the fact that used products, also denoted as cores, are a less homogeneous input resource than traditional raw materials or components. First, used products are dispersed over a potentially large group of users; second, used products may differ in their quality status. CLSCs deal with these particular supply issues through the processes of product acquisition, grading, and disposition decisions (see Guide and Van Wassenhove 2003). Product acquisition refers to the sourcing and procurement of used products; grading reveals the quality of a given core through inspection and testing; the disposition process assigns a core to a specific recovery process and corresponding distribution channel. In this chapter, we discuss managerial issues pertaining to these processes and review corresponding literature.

On the demand side, novel issues in CLSCs concern the development of new markets for recovered products, components, or materials. These issues are discussed in detail in Chapter 8.

Managing the acquisition, grading, and disposition processes involves decisions on multiple levels. Strategic decisions concern the process design and corresponding resources. Strategic issues are addressed in detail in Part I of this book, and we only comment on a few specific aspects. The focus of this chapter is on the tactical and operational planning level. Sections 6.2 through 6.4 address product acquisition, grading, and disposition decisions, respectively. In Section 6.5, we synthesize our discussion, address the interaction between the individual processes, and point out open research questions.

6.2 Product Acquisition

Product acquisition activities represent the supply side of CLSCs, feeding used items (cores) into the system. The management of these acquisition activities varies based on the type of CLSC. In particular, we distinguish between "market-driven" and "waste stream" CLSCs (Guide and Van Wassenhove 2001). In a waste stream system, firms passively accept all returned items, and the focus is simply on processing them at minimum cost. In these cases, the role of product acquisition management is minimal, and the focus is on grading and disposition activities, which are discussed later in this chapter.* This differs from the market-driven approach, where the goal is to close the loop by reintroducing items to the market. In a market-driven CLSC, profit maximization is the objective, and acquisition decisions are a key component of the management of the CLSC. In this section, we address market-driven CLSCs, because this is where active management of product acquisition occurs.

Some of the academic works addressing the strategic aspects of remanufacturing (see Chapter 2) is related to the acquisition process. For example, Ferguson and

---

* Similarly, OEMs receiving used items as customer leases expire (Ferguson et al. 2009) or processing commercial returns (Guide et al. 2006b) typically do not control the inflow of used items.

Toktay (2006) note that some OEMs use acquisition to limit the availability of used products for competitors, and Savaskan and Van Wassenhove (2006) examine the trade-offs between indirect (e.g., retailer based) and direct (e.g., prepaid mailers) acquisition channel structures in the reverse supply chain. As these strategic considerations are addressed elsewhere in this book, our focus in this section is on the tactical aspects of used-product acquisition.

The key considerations in a remanufacturer's management of used-product acquisition are the quality (also called "condition") and the quantity/timing* of acquired items. As mentioned above, if the remanufacturer passively accepts used items (exerts no control over quantity), then the role of acquisition management is minimal. Thus, all acquisition models assume some degree of control over quantity, and the research in the area can be divided into two broad streams based on the degree of control over the quality of acquired items. One stream of work assumes that quality can be influenced by the remanufacturer via pricing decisions. In these cases, the remanufacturer pays a higher price for better-quality units, effectively transferring the process of grading used items to the supplier of the items (e.g., the collector or the consumer). In the CLSC, this implies that grading occurs prior to acquisition by the remanufacturer, a fact that reduces uncertainties and streamlines remanufacturing operations. The second stream of research addresses the case where used item quality cannot be influenced by the remanufacturer. In these cases, items are acquired in unsorted lots, and the grading process does not occur until after the remanufacturer has received the items. For this type of CLSC, the focus is typically on using acquisition lot sizing to enable increased selectivity or reduce the probability of a shortfall of remanufacturable items, taking pricing and quality as exogenous.

We begin with the first research stream, where acquisition pricing is a managerial lever for the remanufacturer. For an independent remanufacturer, this might involve quality-dependent cash payments to end users or collectors. For an OEM, buyback programs or trade-in rebates for existing customers might also be employed. Below we summarize several key papers that address this CLSC setting.

Guide et al. (2003) describe quality-dependent pricing using the case of ReCellular, an independent remanufacturer that obtains presorted used cell phones from several sources, including airtime providers and third-party collectors. ReCellular couples the price paid to the quality of each item, according to well-defined categories, which motivates suppliers to provide more phones for categories with higher prices. The authors emphasize the importance of a remanufacturer's ability to influence both the quality and the quantity of acquired items by offering this quality-dependent price. Assuming that supply is an increasing, twice-differentiable function of price, the authors provide profit-maximizing

---

* For the rest of this discussion, we use the term "quantity" to denote both the quantity of used items acquired and the timing of those acquisitions, given that the two decisions are closely related.

quality-dependent acquisition prices and selling prices for remanufactured items. Karakayali et al. (2007) also assume that used items fall into a number of quality categories, and they model supply of a given category as a linear function of price. Optimal acquisition prices for used items and selling prices for remanufactured parts are provided for several channel structures.

Bakal and Akcali (2006) determine optimal acquisition prices for used vehicles, along with optimal selling prices for remanufactured parts from those vehicles. In their model, quality is defined using a simple threshold—all items above the threshold are remanufacturable, and those below the threshold are scrapped. The probability that each acquired item will be remanufacturable (the "yield") increases with the unit price paid, as does the total supply of used items. They find that the ability to postpone the pricing of remanufactured parts until after the exact yield from the acquired lot is realized always outperforms the case where prices are set simultaneously.

Ray et al. (2005) address the case of trade-in rebates, describing how these can be used to exert control over both the quantity and the quality of used durable goods received back from consumers. The quality of each used item is assumed to be a continuous function of the item's age, and the trade-in rebate could be a constant rebate for all replacement customers or an age-dependent one. The paper provides profit-maximizing new product pricing and trade-in rebate offers for constant, age-dependent, and zero rebates, and defines the market characteristics for which each is optimal.

The second stream of research regarding acquisition management addresses the situation where a remanufacturer acquires unsorted lots of used items and cannot influence the quality distribution of those items. If there is no variation in the quality of the acquired items, then the acquisition decision is focused exclusively on quantity, as in the remanufacturer's lot-sizing problem examined by Atasu and Cetinkaya (2006). When quality variability does exist, the unsorted lots must be graded by the remanufacturer after acquisition. It is typically assumed that an ample supply of used items is available, and lot sizes are determined to effectively manage quality variability. We summarize several key papers that address this CLSC setting in the following text.

Zikopoulos and Tagaras (2007) model variable used-product condition using two categories, where items in one of the categories are not suitable for remanufacturing. Their model includes multiple potential sources of used items, with each source having its own (uncertain) proportion of remanufacturable items. Given that some items cannot be remanufactured, the acquisition quantity might exceed the target production quantity, and increasing acquisition amounts can reduce the probability of a shortfall. The authors optimize acquisition quantities from each collection site and the total production quantity for a firm facing a single uncertain demand.

Galbreth and Blackburn (2006) examine the case where used-product condition can be approximated by a continuum (i.e., there are many different possible conditions). In this environment, higher acquisition amounts enable the firm to be more selective in meeting a given demand, choosing only the best items to remanufacture and scrapping the others. This implies a basic trade-off of acquisition and scrapping costs versus remanufacturing costs, and the authors derive acquisition amounts that optimize this trade-off in a single-period model. Galbreth and Blackburn (to appear) extend this work, using the order statistics of the used items to provide a closed-form expression for the optimal quantity of items to acquire. That paper also presents optimal acquisition quantities for the two-condition case where both categories are assumed to be remanufacturable, but at different costs.

Robotis et al. (2005) also model used-product quality as a continuous variable, where items are acquired from two classes of suppliers in unsorted lots in a single period. Threshold quality levels are used to divide acquired items into two quality classes, each with its own demand distribution and a fixed market price. Remanufacturing can increase the quality of an item to meet a threshold level. Acquisition quantities from the two supplier classes are optimized, along with the quality ranges for which remanufacturing should occur. The paper quantifies the value of the remanufacturing option in this setting, with the primary conclusion that remanufacturing can lead to lower used-product acquisition quantities and higher profits.

Table 6.1 categorizes the product acquisition papers discussed above based on two dimensions—the ability to influence quality through pricing, and the manner in which quality variability is modeled.

Table 6.1   Incorporating Quality Variability in Models of Used Item Acquisition

	Items Are either Remanufacturable or Not	*Multiple Remanufacturable Conditions (Discrete Quality Set)*	*Multiple Remanufacturable Conditions (Continuous Quality)*
Quality is influenced by the price paid	Bakal and Akcali (2006)	Guide et al. (2003), Karakayali et al. (2007)	Ray et al. (2005)
Quality cannot be influenced	Zikopoulos and Tagaras (2007)	Galbreth and Blackburn (to appear)	Robotis et al. (2005), Galbreth and Blackburn (2006, to appear)

6.3 Grading

Used items differ in their quality (condition). In many cases, this quality is not known a priori. In that case, used items have to be evaluated so as to determine their suitability for value recovery. This is the domain of the grading process in CLSCs.

In general, there exists some a priori knowledge of the quality distribution in the available items, and there are some alternatives for the grading process, including the "do nothing" alternative of not testing prior to the disposition decision. Given these inputs, the objective is to determine those grading decisions that optimize the (economic) performance of the relevant system.

It is obvious that the boundaries of the "relevant" system are ambiguous. If the system under examination is strictly the grading system, then the problem is simplified but its optimal solution will only be a local optimum. If the system includes the acquisition or disposition processes and decisions, then it is more complete but more difficult to analyze and optimize globally. The issue of joint grading and acquisition or disposition decisions will be discussed in the concluding section of this chapter. In this section, we concentrate on the grading issues and decisions specifically.

The grading problem may involve decisions at different levels:

■ At the strategic/long-term level, grading is connected to product design and supply-chain design. With regard to product design, the important question is whether the need to assess the quality of used items must be taken into account in the design process. For example, a recent tendency is to implant electronic devices (e.g., chips) in the products with the purpose of recording data, which will allow a quick evaluation of the condition of the item when it is returned for possible remanufacturing, without the need for complete (and expensive) disassembly. With regard to the design of the reverse supply chain, a critical issue is the appropriate location of the grading operations. Should they be performed at the collection sites, at the remanufacturing facility, or at some other location? It must be noted, though, that the latter issue is strategic only to the extent that the respective decision is practically impossible to reverse because of extremely large costs of the initial investment (e.g., expensive, heavy, specialized equipment). If, however, the grading operations are lean and easily transferable then their location becomes a tactical issue.

■ At the tactical/medium-term level, the grading decisions refer to the grading method and the classification scheme. Specific issues that have to be addressed include the detail and the accuracy of the grading scheme: how many quality states/categories will be used, which variable(s) or attribute(s) will be evaluated and how?

■ The operational/short-term level is related to real-time grading decision, based on actual needs (demand) and value recovery capacity. For example, there may be a choice between using a faster (less expensive) grading method, or a slower but more expensive and accurate method. Preference is given to one method or the other depending on the costs and the urgency of the need to satisfy a given order. These decisions are by their nature closely related to disposition issues.

As the main focus of this chapter is on tactical matters, we now direct our attention to decisions about the grading method and the time or location of the grading process. There are several factors that complicate these decisions:

- Multiple quality states of used (collected) items or multiple recovery options
- Uncertain quality distribution of used items
- Limited accuracy of grading, classification errors
- Uncertain quantity (supply) of used items
- Uncertain demand for remanufactured products
- Complex reverse supply chain with multiple collection sites, collection center, etc., resulting in multiple possible alternative grading configurations

These factors have drawn the attention of researchers. In what follows, we review the associated literature. We distinguish two streams of papers. The first stream takes the actual grading decisions as given and focuses on assessing the "value of information" obtained through grading. Thereby, these papers essentially examine the economic viability of a specific grading operation. The second stream encompasses papers that explicitly compare alternative grading options.

We start with the "value of information" stream. The first publication that refers directly to the grading problem in reverse supply chains as delineated here is by Souza et al. (2002), who examine different production planning and control strategies for the case of a remanufacturing facility with returned products that fall into three different quality classes, each requiring a different remanufacturing process. The proportions of used products that fall into each of the three classes are known. The cost of the grading and sorting operation is explicitly not taken into account, as the emphasis is on product mix decisions at the tactical level and dispatching rules at the operational level. The value of quality information subject to grading errors is studied via simulation, assuming that the grading/sorting procedure has a constant probability of product misclassification. The effect of grading errors on system profitability is found to be minor.

Ketzenberg et al. (2003) also examine the effect of advanced (before disassembly) quality information as a side issue in their simulation analysis of a mixed assembly–disassembly line for remanufacturing. The assumptions regarding the grading operation are similar to those of Souza et al. (2002) with two exceptions: (a) there are only two quality classes (recoverable and unrecoverable parts) and (b) sorting is error free. Note that as there are only two quality classes, corresponding to products that can or cannot be remanufactured, the authors use the term "yield information" rather than quality information. The same terminology is also adopted in most of the later papers with the same assumption about quality categories of returns.

Ferrer (2003) studies the value of information about the yield of returns at a remanufacturing facility, examining different scenarios in the context of a single-period lot-sizing problem. Under all scenarios, the demand is assumed to be known and can be satisfied with remanufactured or new items in ample supply. There is no distinct or explicit grading operation; in one scenario, the yield is revealed only after remanufacturing, while in other scenarios the recoverable items are identified after disassembly with complete accuracy. The latter scenarios may be perceived as equivalent to cases where an error-free and costless sorting operation is feasible but only after disassembly. The author compares two such cases, one in which the yield is a random variable and another where the exact recovery yield is known in advance but the actual recoverable cores are identified only after disassembly. Knowing the exact yield in advance allows the determination of the used-items lot size with complete avoidance of shortage and holding (salvage) costs. It is concluded that the benefits of early yield information increase in the variability of the yield and in the acquisition, processing, and holding costs. Ferrer and Ketzenberg (2004) study the multi-period multipart extension of the model of Ferrer (2003) and arrive at similar conclusions.

The value of timely grading and sorting of returns is also examined by Aras et al. (2004) in the context of joint manufacturing and remanufacturing systems. There are two quality classes of returned products: high quality and low quality. Contrary to the models in Ketzenberg et al. (2003), Ferrer (2003), and Ferrer and Ketzenberg (2004), it is assumed that both returned product categories can be successfully remanufactured but at a different cost. The inspection process is explicitly taken into consideration in the continuous-time Markov chain model, but its cost is ignored as irrelevant, because all returns are inspected and graded. The quality categorization (grading) is assumed to be error free. The value of grading is examined by comparing the optimal cost of this system with the cost of a system where the quality of returns is ignored in deciding which items to remanufacture or dispose. It is concluded that the value of grading is higher when the quality difference between the two classes is large, the quality of both return types decreases and the volume of returned products increases.

Guide et al. (2005) evaluate the potential savings from the introduction of error-free testing of used returned notebooks in Hewlett-Packard's reverse supply chain. The test determines which of these items are of sufficiently good quality to undergo only low-touch refurbishment and which items require high-touch refurbishment. The multi-period linear-programming model is simplistic in that demand is taken as known and the proportions of high-quality and low-quality returns are also deterministic but the testing cost is explicitly taken into account. It is shown that if the average incoming quality is high, then the policy with testing outperforms the old no-testing policy, whereby all returned units should go through high-touch refurbishment.

Zikopoulos and Tagaras (2008) study the value of quick but inaccurate grading of returns before disassembly in a simple two-stage system (disassembly–remanufacturing), concentrating mostly on the single-period setting. The used items are assumed to be in ample supply, are procured from a single collection site, and their condition is dichotomous: remanufacturable or non-recoverable. The base model concerns a remanufacturing operation without grading of used products before disassembly and determines the procurement and remanufacturing quantities that maximize the expected profit under quality and demand uncertainty. Then, an alternative system is analyzed, where the remanufacturer has the option to establish a sorting/grading procedure just before the disassembly operation so as to identify the remanufacturable units before the typically expensive dismantling process. This grading operation is subject to classification errors. The two systems are compared in terms of profitability, and the comparison reveals the conditions under which timely grading of returns is economically justifiable. The infinite-horizon problem is examined briefly for the case of a reverse supply chain with a single collection site and stochastic yield of returns.

Behret and Korugan (2009) use simulation to analyze a hybrid manufacturing–remanufacturing system with uncertainties in the quality and timing of returns. All returned products are inspected and classified into three quality levels requiring different remanufacturing efforts and having different but deterministic yields, because not all returns are eventually remanufacturable. The actual condition (remanufacturability) of each item is revealed only during the remanufacturing operation. It is concluded that the quality classification of returned products results in substantial cost savings.

As discussed above, a second stream of papers explicitly compares multiple grading alternatives. Within this stream, Blackburn et al. (2004) are the first to discuss the appropriate location of the grading operation, making a distinction between testing and evaluation of returns at a centralized facility and decentralized testing and evaluation at the points of return; the latter model is termed "preponement." They explain that a reverse supply chain with centralized grading of all returns is efficient in that it exploits economies of scale in processing and transport, while a reverse supply chain with decentralized early grading is more responsive, reduces time delays, and thus improves asset recovery especially for items with quick value erosion. The authors point out that a prerequisite for preponement is technical feasibility of product grading at the collection points with quick and inexpensive methods. They argue that the trade-off between grading efficiency and responsiveness depends primarily on the marginal time value of the product. The article is extensive and descriptive; in Guide et al. (2006b), an analytical model is developed to quantify the relevant trade-offs in the grading location decision.

Similar to Zikopoulos and Tagaras (2008), the paper of Tagaras and Zikopoulos (2008) studies the value of information about the quality (yield) of returns through testing and grading before disassembly but in the context of a richer structure, namely, in a reverse supply chain with one remanufacturing facility and multiple collection sites. The grading operation is again subject to classification errors and may take place either centrally at the remanufacturing facility or locally at the collection sites. If it takes place at the remanufacturing facility, the unit cost of grading is assumed to be lower because grading is generally performed more efficiently centrally or the collection sites charge a premium for grading. A model is developed for the case of deterministic yields at the collection sites, stochastic demand for remanufactured items, and infinite horizon. The model results in the determination of the optimal quantities of returns to be procured from the collection sites for three alternative grading configurations: no grading, centralized grading at the remanufacturing facility, and decentralized grading at the collection sites. The paper also quantifies the value of grading and derives conditions showing when and where quality grading of returns is worthwhile from an economic point of view. The main theme of this paper with regard to the appropriate time and location of the quality grading operation, that is, centralized versus decentralized, is similar to the discussion in Blackburn et al. (2004). However, while in the latter the emphasis is on marginal time value of the product, in the quantitative models of Tagaras and Zikopoulos (2008) the differentiating elements of the two alternatives are the grading costs and the savings in transportation cost due to the avoidance of transporting non-remanufacturables when grading takes place at the collection sites.

Denizel et al. (to appear) examine a remanufacturing environment where known quantities of returned products (cores) are graded and grouped into multiple different quality levels. The cost of grading is explicitly taken into account. Graded cores are remanufactured to meet deterministic nonstationary demand for remanufactured products over multiple time periods. The remanufacturing costs differ across quality grades. Cores can be kept in stock to be graded in the future. Graded cores can be kept in stock to be remanufactured in the future and can also be salvaged at any time. The problem is to determine, in each period, how many of the available cores to grade, how many of the graded cores to remanufacture, and how many to salvage, so as to maximize total expected profit subject to capacity constraints. The problem is formulated as a stochastic program where the outcome of the grading process in each period is a random variable. Among other issues, the paper examines numerically the effect of the grading cost on the firm's profit and the optimal values of the decision variables.

Ferguson et al. (2009) study a production planning problem similar to that of Denizel et al. (to appear) but with uncertain returns and demand for remanufactured products, with and without capacity constraints. The paper examines explicitly the value of a nominal quality grading system without classification errors and the benefits of maintaining separate inventories for each quality grade. More specifically, the quality of each return is represented by a real number q in [0, 1] with a known probability distribution. Returns with quality in $[q_1, 1]$ are remanufacturable and the range $[q_1, 1]$ is divided into N slices so as to classify remanufacturable returns into N quality grades. However, the holding costs, remanufacturing costs, and salvage values are functions of q. The numerical investigation shows that the

Table 6.2   Grading-Related Literature

	Deterministic Yield	*Stochastic Yield*
Value of information of given grading system	Souza et al. (2002), Ketzenberg et al. (2003), Aras et al. (2003), Guide et al. (2005), Behret and Korugan (2009)	Ferrer (2003), Ferrer and Ketzenberg (2004), Zikopoulos and Tagaras (2008)
Comparison of multiple grading options	Blackburn et al. (2004), Tagaras and Zikopoulos (2008), Ferguson et al. (2009)	Denizel et al. (to appear)

grading system increases profit by an average of 4 percent over a wide range of realistic parameters.

Table 6.2 summarizes the papers discussed in this section. In addition to the two streams distinguished above, the following characteristics of the grading-related literature can be observed:

- The quality state of returns is typically treated as a discrete variable. The usual assumption is that there exist two or three quality classes. In some papers, all returns are assumed to be recoverable but with remanufacturing cost depending on the quality level, while in other papers a proportion of the returned products is non-recoverable.
- The grading yield is assumed to be known in most cases, with the exceptions of Ferrer (2003), Ferrer and Ketzenberg (2004), Zikopoulos and Tagaras (2008), and Denizel et al. (to appear), where yield is treated as a random variable (see Table 6.2).
- The cost of the grading operation is modeled explicitly in Guide et al. (2005), Zikopoulos and Tagaras (2008), Tagaras and Zikopoulos (2008), Denizel et al. (to appear), and Ferguson et al. (2009).
- The grading operation is assumed to be error-free in all models except for Souza et al. (2002), Zikopoulos and Tagaras (2008), and Tagaras and Zikopoulos (2008).
- Almost all papers focus on a single remanufacturing facility, with or without parallel manufacturing of new products. Blackburn et al. (2004), Guide et al. (2006b), and Tagaras and Zikopoulos (2008) are the only papers that examine a supply chain with a central remanufacturing or testing facility and multiple collection sites, where testing and sorting are also feasible.
- All papers discussed here assume a single recovery option—remanufacturing—for recoverable returns. Other research that exploits multiple recovery options (e.g., dismantling for spare parts) is discussed in the next section.

6.4 Disposition Decisions

CLSCs, in general, include multiple options regarding the further treatment of the acquired cores. In the simplest case, the choice is between some valuable recovery option, such as remanufacturing, and disposal; this case was covered in the previous section. Other cases also include multiple recovery alternatives, on a product, component, or material level (Thierry et al. 1995). For example, used computer equipment may be refurbished and resold, dismantled to obtain valuable spare parts, or recycled for its precious metal content (Fleischmann et al., 2005). The availability of multiple recovery options raises the question of which option to choose for a given core. This is the domain of the disposition decision.

The disposition problem can be defined as follows: given a set of acquired cores and a set of available recovery options, find an optimal assignment of cores to recovery options. Typically, the optimality criterion is profitability, which includes revenues, processing costs, inventory costs, and penalty costs. However, the disposition decision could also consider environmental performance metrics.

The disposition problem involves managerial decisions at different planning levels:

- *Strategic disposition decisions* notably concern the process design: When, where, and based on which information is the core assignment made? This involves the usual trade-offs between centralization and decentralization and between responsiveness and efficiency.
- *Tactical disposition decisions* determine planned allocated volumes, based on forecasted acquisition volumes and demand for recovered products. To some extent, these decisions are a mirror image of traditional aggregated production planning. While aggregated production planning seeks the optimal sources to satisfy given (forecast) demand, the disposition decision seeks the optimal use for a given (forecasted) supply of cores.
- *Operational disposition decisions* concern the actual assignment of a specific, given core. Disposition decisions on this level bear similarities with revenue management, in the sense that they seek to maximize the returns generated with a limited set of resources (i.e., cores).

The disposition problem is easy for a single core and complete transparency regarding all recovery options—simply pick the option with the highest marginal profit. In reality, however, the right disposition decision often is much less obvious. Companies are faced with several factors that complicate the matter, including the following.

Demand uncertainty. The actual demand for a given recovery option often is unknown at the time of the disposition decision. Demand uncertainty tends to be relatively high for recovered products, which are often less well established than traditional new products. As most recovery options require a certain amount of processing, the disposition decision is irreversible to some extent, and demand uncertainty makes it a risky decision. Thus, companies have to make a risk-return trade-off between the up-front processing costs of each recovery option and the expected returns. Ignoring demand uncertainty tends to bias the disposition decision toward "high end" options such as refurbishing or remanufacturing and may forego margins from cheaper but safer alternatives, such as parts harvesting. As in traditional supply chains, one way to deal with demand uncertainty is to postpone the disposition decision and to process cores on demand. However, this comes at the expense of an investment into responsive processing.

Uncertain quality. One of the main distinctions between CLSCs and traditional supply chains is the degree of supply uncertainty. Used products are a much less homogeneous input resource than conventional raw materials or components. The grading operation discussed in the previous section seeks to resolve this quality uncertainty. There is a trade-off between the grading effort and the quality information available for the disposition decision. Core quality may affect both the processing requirements and the processing yield. In general, high-end recovery options are more quality dependent. Thus, the disposition decision faces a risk-return trade-off again.

Uncertain acquisition volumes. Not only the quality but also the amount of cores that a company will be able to acquire against a given price is uncertain, in general. This amount depends on many factors, such as the number of products in use, their life cycle, and their original selling date. The uncertain acquisition volume affects the disposition decision through the opportunity costs. Assigning a core to one recovery option also means withholding it from other options, thus entailing opportunity costs. These opportunity costs increase with decreasing future acquisition volumes.

Structural fluctuations of supply and demand. Uncertainty is not the only complicating factor in the disposition decision. Predictable fluctuations in both supply and demand volumes also add to the complexity of the problem. If supply and demand move in an asynchronous way, which is not unusual as high demand increases the competition for cores, the disposition decision has to consider a longer planning horizon and make a trade-off between maximizing the current contribution of a core and holding the core in inventory for future opportunities.

Varying capacity utilization. Related to the previous issue, fluctuating supply volumes and quality also result in varying capacity utilization and throughput times. This, in turn, may result in nonstationary unit processing costs and possibly revenues, which further complicates the relevant trade-offs.

Interrelation between recovery options. Another difficulty stems from the fact that different recovery options may be interdependent. For example, disassembling (or shredding) a core yields multiple parts (or materials) simultaneously, which may have different demand volumes. A good disposition decision has to consider all of these parts (or materials) jointly and seek a global optimum.

Clearly, a simple disposition decision merely based on unit margins is not capturing these issues appropriately. Several researchers have proposed more advanced approaches to the disposition decisions in CLSCs. In the remainder of this section, we review analytical models available in the literature. Table 6.3 summarizes their positioning within the earlier discussion.

We are aware of two papers that analyze the impact of the disposition process design (Guide et al. 2005, Guide et al. 2006b). Both papers focus on commercial product returns (also known as "consumer returns") and investigate the potential benefits of shifting from a centralized to a decentralized grading and disposition decision, a shift known as "preponement" (Blackburn et al. 2004, Guide et al. 2006b; see also Section 6.3). While centralization exploits economies of scale, a decentralized disposition decision close to the source enhances supply-chain responsiveness. This is particularly relevant for commercial returns that depreciate quickly, such as electronic equipment.

Guide et al. (2005) distinguish between light internal refurbishment, more substantial external refurbishment, and unprocessed broker sales. They propose a multi-period network-flow model to determine the product quantities assigned to each of these options, for given supply volumes. Guide et al. (2006b) consider two disposition options, namely, restocking for the primary market and remanufacturing for secondary sales. The authors develop queuing-network approximations for the throughput times of product returns. They then compare different process configurations, based on the price decay associated with these throughput times.

On a tactical planning level, Kleber et al. (2002) focus on the impact of non-stationary acquisition and demand volumes. Considering an arbitrary number of disposition options, they determine a dynamic allocation policy, based on an optimal-control model. The policy dynamically builds up and consumes inventory in response to supply and demand fluctuations.

Another stream of research focuses on the interdependencies between different input products as well as different product components. In this approach, the disposition options typically correspond to varying levels of product disassembly. Krikke et al. (1998) propose a mixed integer linear program (MILP) model for choosing disassembly strategies for multiple products simultaneously, so as to meet overall financial or environmental targets. Spengler et al. (2003) analyze a related short-term planning problem. Their network flow model determines the daily recycling flows of a scrap processor. In addition, the scrap acquisition volumes are also determined. A similar network flow model is proposed by Jorjani et al. (2004).

Another set of short-term planning models focuses on the impact of supply and demand uncertainty. The methodology builds on stochastic inventory control. In this line of research, Inderfurth et al. (2001) consider the allocation of randomly arriving cores to multiple disposition options facing stochastic demand. Assuming

Table 6.3 Disposition Decision Models

	Demand Uncertainty	Uncertain Core Quality	Uncertain Future Acquisition	Known Supply and Demand Fluctuations	Varying Capacity Utilization	Interdependent Recovery Options
Strategic disposition decisions						
Guide et al. (2005)		X		X		
Guide et al. (2006b)		X	X		X	
Tactical disposition decisions						
Kleber et al. (2002)				X		
Krikke et al. (1998)						X
Operational disposition decisions						
Spengler et al. (2003)						X
Jorjani et al. (2004)						X
Inderfurth et al. (2001)	X		X			
Ferguson et al. (2008)	X		X			
Karaer and Lee (2007)	X	X	X			
Guide et al. (2006a)		X	X		X	

a linear allocation of shortages, they minimize expected inventory holding costs. Ferguson et al. (2008) allow an arbitrary allocation of available cores to either refurbishing or parts salvaging and maximize expected contribution margins. They highlight the analogy between this problem and traditional revenue management. Karaer and Lee (2007) consider disposal, direct reselling, and remanufacturing, with the latter two acting as substitutes for new procurement. The disposition decision is determined by the core quality, which is stochastic. Guide et al. (2006a) use a queuing approach to investigate the impact of uncertain core supply and quality. The disposition decision chooses between internal remanufacturing and external material recycling and is dependent on the core quality, which is revealed in the grading step. A numerical analysis shows this policy to outperform a disposition decision based on the queue length at the remanufacturing facility.

6.5 Conclusions

In this chapter, we have discussed key inbound processes of CLSCs, namely, product acquisition, grading, and disposition. These processes are critical to the management of CLSCs given that used products are a less-standardized input resource than traditional raw materials or components. In our discussion, we have emphasized the economic role of a CLSC as a broker between customers disposing of used products and customers acquiring recovered products. To fulfill this role, the inbound processes discussed in this chapter have to be complemented with appropriate outbound processes, such as remarketing and redistribution. These are discussed in Chapter 8.

To summarize our literature review of this area, product acquisition models focus on decisions concerning inbound quantities and their timing, and, potentially, also the management of inbound product quality. Grading models primarily assess the value of information regarding product quality. Only a few models compare different grading alternatives. Regarding the disposition decision, separate streams of the literature focus on the impact of uncertainty (in demand, supply, or quality), nonstationary supply and demand, and interaction between multiple products or components.

A few general observations regarding this literature are worth highlighting. First, the number of available papers on acquisition, grading, and disposition is fairly small. Second, these papers are rather recent, even relative to the field of CLSCs, which in itself is still a young topic; almost all papers have appeared within the last ten years, most of them within the last five. Thus, acquisition, grading, and disposition issues appear to have attracted research interest only recently. This is remarkable given that these are key processes that distinguish CLSCs from conventional supply chains. Given this state of the literature, many open research opportunities remain in each of the three areas discussed in this chapter. As discussed in the preceding sections, several papers are available on specific issues. However, a coherent body of literature is yet to emerge. To guide this process and to assure reasonable modeling assumptions, published case studies would be highly valuable in this field.

Based on our literature reviews, we suggest a few potential directions in each of the three areas considered.

Regarding product acquisition, more research is needed into the appropriate way to model used item availability. Although used items are often in ample supply, there are cases in which supply is limited. As suggested by Guide and Jayaraman (2000) nearly a decade ago, models to better forecast the quantity of used items available, particularly incorporating product life cycle considerations, are needed. Ideally, these models would differentiate between the availability of different quality levels at different prices. In addition, the impact of legislation, including disposal fees, recycling subsidies, etc., on used item supply is not well understood (Wojanowski et al. 2007). Finally, a better understanding of the sequence of acquisition vis-à-vis grading is needed, that is, determining the conditions under which it makes sense to acquire ungraded lots versus transferring the grading process to the collector and acquiring graded items.

Regarding the grading-oriented literature, most of the current papers focus on the value of information in a given system. We see a clear need for additional studies comparing different grading options. These should include richer supply-chain structures, for example, separate testing facilities either for the entire chain or for a group of collection sites. This would allow a deeper examination of the optimal timing and the location of used-product quality grading. Similarly, different grading methods should be compared, such as inexpensive and inaccurate versus expensive and more accurate methods. Yet another promising direction concerns the comparison between ex-post grading and continuous monitoring of the product during the usage phase.

In the area of used-product disposition, a deeper understanding of the impact of uncertainty and the interaction between the different sources of uncertainty would be valuable in our opinion. For example, different recovery options serve different markets with different degrees of demand uncertainty. How should these different demand risks be reflected in the disposition decision? Another open issue concerns the fair valuation of different recovery alternatives. Many companies struggle with the accounting of used products as resources. This has an immediate bearing on the disposition decision. Artificial book values of used products may lead to a significant distortion of the disposition process.

In addition to open questions regarding the management of the individual processes, important issues arise from their coordination. Most of the currently available papers focus primarily on one of the three subprocesses. However, acquisition, grading, and disposition are strongly interdependent. For example, the appropriate acquisition volumes depend on the market potential of the used products, which itself depends on the product quality; the value of information for grading is driven by the fact that it enables either a better acquisition decision or a better disposition decision; products can be graded prior to or after acquisition; the disposition process can only allocate available products, and these are a consequence of the acquisition process; the disposition decision depends on opportunity costs, which depend on future acquisition volumes. The systematic analysis of these interdependencies and of corresponding coordination mechanisms opens a rich field for meaningful future CLSC research.

References

Aras, N., T. Boyaci, and V. Verter (2004). The effect of categorizing returned products in remanufacturing. *IIE Transactions* 36(4): 319–331.

Atasu, A. and S. Cetinkaya (2006). Lot sizing for optimal collection and use of remanufacturable returns over a finite life-cycle. *Production and Operations Management* 15(4): 473–487.

Bakal, I.S. and E. Akcali (2006). Effects of random yield in remanufacturing with price-sensitive supply and demand. *Production and Operations Management* 15(3): 407–420.

Behret, H. and A. Korugan (2009). Performance analysis of a hybrid system under quality impact of returns. *Computers and Industrial Engineering* 56(2): 507–520.

Blackburn, J.D., V.D.R. Guide Jr., G.C. Souza, and L.N. Van Wassenhove (2004). Reverse supply chains for commercial returns. *California Management Review* 46(2): 6–22.

Denizel, M., M. Ferguson, and G.C. Souza (to appear). Multi-period remanufacturing planning with uncertain quality of inputs. *IEEE Transactions on Engineering Management*.

Ferguson, M.E. and L.B. Toktay (2006). The effect of competition on recovery strategies. *Production and Operations Management* 15(3): 351–368.

Ferguson, M.E., M. Fleischmann, and G.C. Souza (2008). Applying Revenue Management to the Reverse Supply Chain. *Working Paper*. Rotterdam School of Management, Erasmus University, Rotterdam, the Netherlands.

Ferguson, M., V.D. Guide Jr., E. Koca, and G. Souza (2009). The value of quality grading in remanufacturing. *Production and Operations Management* 18(3): 300–314.

Ferrer, G. (2003). Yield information and supplier responsiveness in remanufacturing operations. *European Journal of Operational Research* 149(3): 540–556.

Ferrer, G. and M.E. Ketzenberg (2004). Value of information in remanufacturing complex products. *IIE Transactions* 36(3): 265–277.

Fleischmann, M., J.A.E.E. van Nunen, B. Gräve, and R. Gapp (2005), Reverse logistics – capturing value in the extended supply chain. In: An, C. and H. Fromm (eds.), *Supply Chain Management on Demand*. Springer, Berlin, Germany, pp. 167–186.

Galbreth, M.R. and J.D. Blackburn (2006). Optimal acquisition and sorting policies for remanufacturing. *Production and Operations Management* 15(3): 384–392.

Galbreth, M.R. and J.D. Blackburn (to appear). Optimal acquisition quantities in remanufacturing with condition uncertainty. *Production and Operations Management*.

Guide Jr., V.D.R. and V. Jayaraman (2000). Product acquisition management: Current industry practice and a proposed framework. *International Journal of Production Research* 38(16): 3779–3800.

Guide Jr., V.D.R. and L.N. Van Wassenhove (2001). Managing product returns for remanufacturing. *Production and Operations Management* 10(2): 142–155.

Guide Jr., V.D.R. and L.N. Van Wassenhove (eds.) (2003). *Business Aspects of Closed-Loop Supply Chains.* Carnegie Mellon University Press, Pittsburgh, PA.

Guide Jr., V.D.R., R. Teunter, and L.N. Van Wassenhove (2003). Matching demand and supply to maximize profits from remanufacturing. *Manufacturing & Service Operations Management* 5(4): 303–316.

Guide Jr., V.D.R., L. Muyldermans, and L.N. Van Wassenhove (2005). Hewlett-Packard company unlocks the value potential from time-sensitive returns. *Interfaces* 35(4): 281–293.

Guide Jr., V.D.R., E. Gunes, G.C. Souza, and L.N. Van Wassenhove (2006a). The Optimal Disposition Decision for Product Returns. *Working Paper.* Robert Smith School of Business, University of Maryland, College Park, MD.

Guide Jr., V.D.R., G.C. Souza, L.N. Van Wassenhove, and J.D. Blackburn (2006b). Time value of commercial product returns. *Management Science* 52(8): 1200–1214.

Inderfurth, K., A.G. de Kok, and S.D.P. Flapper (2001). Product recovery in stochastic remanufacturing systems with multiple reuse options. *European Journal of Operational Research* 133(1): 130–152.

Jorjani, S., J. Leu, and C. Scott (2004). Model for the allocation of electronics components to reuse options. *International Journal of Production Research* 42(6): 1131–1145.

Karaer, O. and H.L. Lee (2007). Managing the reverse channel with RFID-enabled negative demand information. *Production and Operations Management* 16: 625–645.

Karakayali, I., H. Emir-Farinas, and E. Akcali (2007). An analysis of decentralized collection and processing of end-of-life products. *Journal of Operations Management* 25(6): 1161–1183.

Ketzenberg, M.E., G.C. Souza, and Guide Jr., V.D.R. (2003) Mixed assembly and disassembly operations for remanufacturing. *Production and Operations Management* 12(3): 320–335.

Kleber, R., S. Minner, and G. Kiesmüller (2002). A continuous time inventory model for a product recovery system with multiple options. *International Journal of Production Economics* 79(2): 121–141.

Krikke, H.R., A. van Harten, and B.C. Schuur (1998). Mixed policies for recovery and disposal of multiple-type consumer products. *Journal of Environmental Engineering-ASCE* 124(4): 368–379.

Ray, S., T. Boyaci, and N. Aras (2005). Optimal prices and trade-in rebates for durable, remanufacturable products. *Manufacturing & Service Operations Management* 7(3): 208–228.

Robotis A., S. Bhattacharya, and L.N. Van Wassenhove (2005). The effect of remanufacturing on procurement decisions for resellers in secondary markets. *European Journal of Operational Research* 163(3): 688–705.

Savaskan, R.C. and L.N. Van Wassenhove (2006). Reverse channel design: The case of competing retailers. *Management Science* 52(1): 1–14.

Souza, G.C., M.E. Ketzenberg, and Guide Jr., V.D.R. (2002). Capacitated remanufacturing with service level constraints. *Production and Operations Management* 11(2): 231–248.

Spengler, T., M. Ploog, and M. Schröter (2003). Integrated planning of acquisition, disassembly and bulk recycling: A case study on electronic scrap recovery. *OR Spectrum* 25(3): 413–442.

Tagaras, G. and C. Zikopoulos, C. (2008). Optimal location and value of timely sorting of used items in a remanufacturing supply chain with multiple collection sites. *International Journal of Production Economics* 115(2): 424–432.

Thierry, M.C., M. Salomon, J.A.E.E. van Nunen, and L.N. Van Wassenhove (1995). Strategic issues in product recovery management. *California Management Review* 37(2): 114–135.

Wojanowski, R., V. Verter, and T. Boyaci (2007). Retail-collection network design under deposit-refund. *Computers & Operations Research* 34(2): 324–345.

Zikopoulos, C. and G. Tagaras (2007). Impact of uncertainty in the quality of returns on the profitability of a single-period refurbishing operation. *European Journal of Operational Research* 182(1): 205–225.

Zikopoulos, C. and G. Tagaras (2008). On the attractiveness of sorting before disassembly in remanufacturing. *IIE Transactions* 40(3): 313–323.

Chapter 7

Production Planning and Control for Remanufacturing

Gilvan C. Souza

Contents

7.1 Introduction

Chapter 6 provided a thorough review of the academic literature on product acquisition, grading, and disposition in closed-loop supply chains (CLSCs). In this chapter, we focus on a key disposition decision—remanufacturing—as it has the potential to be the most profitable among other disposition decisions such as dismantling for spare parts and recycling. In this chapter, we make a fundamental assumption: that remanufacturing is indeed the most attractive disposition decision to the firm on a unit margin basis. This means that the firm would always prefer to remanufacture a (good-quality) return than to use an alternative disposition decision, given enough capacity. As a result, we use the convention that returns that are not used for remanufacturing can be salvaged throughout this chapter; salvaging a return means using an alternative (less profitable on a unit margin) disposition decision.

As discussed extensively in Chapter 6, production planning for remanufacturing is different from production planning for new products because the basic material input for remanufacturing—cores or returns—is not homogeneous; there are differences in their quality and availability during the planning horizon. A good-quality return demands less processing capacity from the facility, and costs less to remanufacture than a bad-quality return. Thus, if the firm has excess returns in a given period, it can salvage them (e.g., by selling to a recycler, or dismantling for spare parts), or keep them in inventory for future use. Demand forecasts for remanufactured products is nonstationary (i.e., varies from period to period) throughout the planning horizon, and there are remanufacturing capacity constraints that can also be time-varying. In this chapter, we provide methodologies for planning remanufacturing in such an environment. For more information on the environment faced by remanufacturing firms when planning production, please see Guide (2000) and Souza (2008).

Given that the academic literature on the subject has been reviewed in Chapter 6, we hereby provide two models that firms can use as a decision support when planning their remanufacturing operations. The first model is an optimization model that requires the use of an optimization software (e.g., Excel Solver), and it is built around traditional aggregate planning (also known as sales and operations planning) optimization models for forward chains. The model can be implemented in a spreadsheet, although it requires a level of detail in data that may preclude its implementation for some remanufacturers. In contrast, the second model is based on standard MRP (materials requirement planning) logic, which is easily implementable using a spreadsheet by any practitioner with a practical understanding of MRP, and requires less data; it can also be implemented at the product type (not family) level, although the plan resulting from the optimization must be checked for feasibility given capacity constraints.

To motivate the models, we consider the example of Pitney Bowes (Figure 7.1). Pitney Bowes is an original equipment manufacturer (OEM) based in Stamford, CT, that manufactures mailing equipment that matches customized documents to envelopes, weighs the parcel, prints the postage, and sorts mail by zip code. Pitney Bowes leases about 90 percent of its new product manufacturing, and sells the remainder. A typical leasing contract is for four years, and a typical life cycle for a Pitney Bowes product is six years. At the end of a leasing contract, customers often upgrade to newer generation equipment if it is available. In these cases, the customer returns the used equipment to Pitney Bowes, which tests and evaluates the condition of the used machine, and makes a disposition decision: scrap for

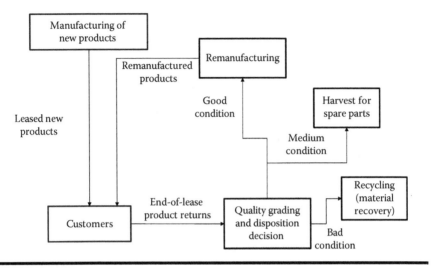

Figure 7.1   Pitney Bowes closed-loop supply chain with remanufacturing.

materials recovery (recycling), which is done for the worst-quality returns; dismantle for spare parts harvesting, which is done for medium-quality returns; or (potential) remanufacturing, which is done for the best-quality returns. Remanufacturing consists of bringing the used product to a common operating and aesthetic standard, often with upgrades in some of the product's functions and the replacement of the wearable parts. Not all returned units designated for remanufacturing are actually remanufactured, as the amount depends on the demand for remanufactured units, which are sold as an (cheaper) alternative to new units. The models presented in this chapter were developed to provide a production plan for this environment.

7.2 Optimization Model for Production Planning

In this section, we present an optimization model for planning remanufacturing. Although the model was inspired by the operations of a particular firm (Pitney Bowes), it is general enough that it can be used by most remanufacturing firms. The model is meant to provide an aggregate production plan, where the planner decides upon the overall remanufacturing quantities for a given product family (e.g., Motorola cell phones), as opposed to individual product models (e.g., the models V750, Razr2, and E8 by Motorola). The model can be extended to planning production at the individual product model level, although the level of detail in the required data for the optimization problem (and the quality of the forecasts) may become a significant issue in its implementation. The planning horizon is divided into T periods of equal length. For aggregate planning, an appropriate period is typically one month and an appropriate planning horizon is typically one year ($T = 12$).

We assume that there are demand forecasts D_t for the family of remanufactured products under consideration for each period t in the planning horizon. The firm remanufactures returns (or cores) that arrive to the facility in different qualities. We also assume that incoming returns can be categorized into G quality grades. Categorization can be made based on visual inspection, reading some counter that tracks product usage (e.g., the number of cycles in copiers), preliminary testing of the different modules in the product, or by a combination of these options. The simplest categorization is into two grades ($G=2$): good and bad. Ferguson et al. (2009) shows that significant cost savings in remanufacturing planning can be achieved with three grades ($G=3$), and there are essentially no benefits in categorizing returns into more than five grades ($G=5$).

Denote $i=1$ as the best-quality grade and $i=G$ as the worst-quality grade. Denote the forecast for the number of returns of each quality i for each period t in the planning horizon by B_{it}; these can be obtained as follows. There are known mechanisms for forecasting the total quantity of returns received in each period (denoted by $B_{\bullet t} = \sum_i B_{it}$). For example, if the firm leases its new item production (such as Pitney Bowes or Xerox), then $B_{\bullet t}$ would be the number of expiring leases in period t. Otherwise, the firm can employ time series methods to forecast $B_{\bullet t}$. Given $B_{\bullet t}$, an estimate of B_{it} would be $B_{it} = q_i B_{\bullet t}$, where q_i is the historic average proportion of returns of quality i, which according to our experience tends to be relatively constant.

A return of quality i has a unit remanufacturing cost (materials and labor) equal to c_i, it consumes a_i hours of remanufacturing capacity, and has a salvage value (i.e., if not remanufactured or kept in inventory) of s_i. As a result of our convention that $i=1$ represents the best-quality return, then $c_1 < c_2 < \cdots < c_G$, $a_1 < a_2 < \cdots < a_G$, and $s_1 \geq s_2 \geq \cdots \geq s_G$. The salvage value for a return is meant to represent the revenue obtained if it is used in an alternative disposition outlet such as dismantling for spare parts or recycling. A higher-quality return is likely to have a higher salvage value, as more parts can be successfully obtained from its dismantling than from a lower-quality return. We also assume that a return of quality i has a unit holding cost of h_i if it is carried in inventory from one period to the next. Holding cost typically includes allocated warehousing and insurance cost, as well as the "interest" on any allocated acquisition cost. For returns coming off of lease, there is usually no direct acquisition cost, whereas for firms that buy returns using a price mechanism (e.g., cell phone remanufacturers such as ReCellular), there are direct acquisition costs.

At each period t, the firm has K_t hours of remanufacturing capacity. This number is basically determined from the bottleneck work center (or resource). If there are multiple bottleneck work centers, then the model formulation can be easily changed to accommodate this environment: return of quality i consumes a_{ij} hours of capacity at work center j, and there are K_{tj} hours of remanufacturing capacity available at work center j. In each period t, the firm receives B_{it} returns of each quality i, and then decides upon the quantity to remanufacture (z_{it}), to salvage (v_{it}), and to carry in inventory to the next period u_{it}. Remanufactured products are used to meet demand in a period, or can be carried in inventory for future periods (this can be necessary if capacity constraints are significant, and there is significant seasonality in demand forecasts).

We summarize our complete notation below, before introducing the optimization model:

Indexes

    i: quality, $i = 1$ (best), ..., G (worse)

    t: time period, $t = 1$, ..., T

Parameters

    D_t: demand forecast for remanufactured products at period t

    B_{it}: quantity of returns of quality i at period t

    s_i: unit salvage value for returns of quality i that are not remanufactured

    h_i: unit holding cost for return of quality i

    h_r: unit holding cost for remanufactured product

    b: unit backlogging cost

    c_i: unit remanufacturing cost for return of quality i

    a_i: unit capacity usage by return of quality i

    K_t: total capacity available at time t

Decision variables

    z_{it}: quantity of quality i returns remanufactured in period t

    v_{it}: quantity of quality i returns salvaged in period t

    y_t^+: inventory of remanufactured products at the end of period t

    y_t^-: backlog of remanufactured products at the end of period t

    u_{it}: inventory of quality i returns at the end of period t

The remanufacturing problem can be formulated as a linear program (LP) as follows:

$$\min TC = \sum_{t=1}^{T} \sum_{i=1}^{G} \left\{ c_i z_{it} - s_i v_{it} + h_i u_{it} \right\} + h_r y_t^+ + b y_t^- \tag{7.1}$$

$$\text{s.t.} \quad \left(y_{t-1}^+ - y_{t-1}^- \right) + \sum_i z_{it} - \left(y_t^+ - y_t^- \right) = D_t, \quad t = 1, \ldots, T, \tag{7.2}$$

$$z_{it} + u_{it} - u_{i,t-1} + v_{it} = B_{it}, \quad i = 1, \ldots, G; t = 1, \ldots, T, \tag{7.3}$$

$$\sum_i a_i z_{it} \leq K_t \quad t = 1, \ldots, T \tag{7.4}$$

$$z_{it}, u_{it}, v_{it}, y_t^+, y_t^- \geq 0, \quad i = 1, \ldots, G; t = 1, \ldots, T. \tag{7.5}$$

The objective function in (7.1) minimizes total costs and comprises variable production cost, salvage cost (note the negative sign for the salvage revenue), holding cost for returns, holding cost for remanufactured products, and backlogging cost. The set of constraints (7.2) represents the inventory balance equations for remanufactured products: for each period t, beginning inventory net of backlogs $(y_{t-1}^+ - y_{t-1}^-)$ plus remanufacturing production $\left(\sum_i z_{it} \right)$ minus demand (D_t) is equal to ending inventory net of backlogs $y_t^+ - y_t^-$. The set of constraints (7.3) represents the inventory balance equations for returns according to quality level: for each quality i and time period t, beginning inventory $(u_{i,t-1})$ plus returns received (B_{it}) minus quantity remanufactured (z_{it}) minus quantity salvaged (v_{it}) equals ending inventory u_{it}. Set (7.4) represents the (aggregate) capacity constraints for each period; if there are multiple (say M) bottleneck work centers as discussed before, then each period t has M capacity constraints of the form $\sum_i a_{ij} z_{it} \le K_{tj}, j = 1, \ldots, M$. Finally, set (7.5) represents the nonnegativity constraints. Our model does not incorporate fixed costs for remanufacturing in a period (also called setup costs) because remanufacturing is for the most part a labor-intensive operation that does not require elaborate machine setups, and our model is also targeted at the family level where setup costs are less of a concern.

The problems (7.1) through (7.5) can be solved numerically to find the optimal production plan. Ferguson et al. (2009) show that if there are no capacity constraints (i.e., K_t is large enough for all t), then the firm always remanufactures the exact quantity demanded in each period if $h_r > h_i$. That is, the ending inventory of remanufactured products y_t^+ is always zero in each period. In essence, the firm carries inventory of remanufactured products only in situations where there are large demand spikes (i.e., D_t is large for some t) coupled with the inability of meeting the exact demand in each period due to capacity constraints.

This formulation assumes that there is no uncertainty in demand or returns; in practice both are likely to occur. There are two ways to deal with this problem. First, the firm may carry some safety stock (SS) of remanufactured products to protect against uncertainty. For example, one can multiply all values of D_t by $(1 + k\sigma)$, where k is a safety factor, and σ is the standard deviation of the ratio of actual demand to forecast (A_t/D_t). This is clearly a heuristic, but one that is easy to implement and use in practice. Second, the firm solves the problems (7.1) through (7.5) on a rolling horizon basis: start at the beginning of the planning horizon (period 1), solve the problem, and implement the first-period solution (z_{i1}, v_{i1}). At the end of period 1, after actual demand A_1 and return quantities are realized, the firm updates the inventory positions and demand forecasts, and resolves the problem for the planning horizon $2, 3, \ldots, T + 1$; this process continues each period. By using a rolling horizon approach coupled with some level of SS, the firm protects itself against forecast error; this is clearly a heuristic solution, however. These two heuristics form the basis for the MRP approach of Section 7.3.

7.3 An MRP Logic to Production Planning*

In this section, we propose a method for production planning that takes into account demand forecast uncertainty, multiple quality grades, and a rolling horizon for updating inventories. This method can be applied to a particular model of a particular product, as long as the firm has forecasts of demand and returns for a planning horizon. To describe the model, we will use an (actual) example from a firm that remanufactures its own products, although the magnitude of the numbers has been disguised.

Consider a firm that is planning remanufacturing for one of its products, K5R. It is the end of September of 2008, and the firm wants to know how many units it should remanufacture in October 2008, as well as beyond. It has kept detailed records of forecasted (F) and actual demand (A) for K5R for previous years. In particular, data for the previous two years is shown in Table 7.1, along with the ratio A/F for each month (a measure of forecast uncertainty and a critical input to our model), and mean and standard deviation of the (A/F) ratios for the 24 months (given by Excel's formulas = AVG(range) and = STDEV(range), respectively).

From Table 7.1, the average and standard deviation of the A/F ratios are 1.0024 and 0.1024, respectively. As the average value of A/F is around 1, the forecast is assumed to be unbiased, that is, the firm is neither (consistently) underestimating nor overestimating the demand. Further, the standard deviation of the A/F ratios, denoted by $\sigma_{A/F}$, is 0.1024, and it indicates that there is forecast uncertainty of about 10 percent around the mean. These two numbers are used to create "fluctuating" SSs for future periods as follows. First, the firm computes an SS multiplier, given by

$$\text{SS multiplier} = 1 + k \cdot \sigma_{A/F}, \tag{7.6}$$

where k is a safety factor that can be computed from underage and overage costs as follows. Denote the (unit) underage cost of not meeting demand for remanufactured products in a period as C_u. If demand not met in a period is lost (i.e., the customer walks away from the deal), then C_u is equal to the remanufactured product's profit margin plus a goodwill cost. If, however, the order is backlogged, then C_u could be a discount the firm has to give to keep the customer happy for having his or her order backlogged. Further, denote the (unit) overage cost of having excess stock of remanufactured products as C_o. In most cases, C_o would simply be the holding cost of keeping the remanufactured product in stock for one period (i.e., warehousing, insurance and interest on capital invested). Then, k can be computed in Excel through the following formula:

---

* This section is based on a joint work with Dan Guide and Mark Ferguson for a remanufacturing firm.

Table 7.1   Forecast and Actual Demand for Product K5R

Year	Month	Forecast (F)	Actual (A)	A/F Ratio
2006	10	225	207	0.922
	11	250	254	1.015
	12	325	360	1.107
2007	1	335	360	1.074
	2	380	410	1.080
	3	475	492	1.035
	4	395	321	0.812
	5	385	363	0.943
	6	495	550	1.111
	7	360	341	0.948
	8	215	230	1.070
	9	310	313	1.011
	10	225	225	1.001
	11	250	273	1.090
	12	325	361	1.111
2008	1	335	407	1.215
	2	380	392	1.032
	3	475	390	0.821
	4	395	380	0.962
	5	385	326	0.848
	6	495	530	1.071
	7	360	354	0.983
	8	215	187	0.868
	9	310	288	0.929
			AVG	1.0024
			STDEV	0.1024

$$k = \text{TINV}\left(2*\left(1 - \frac{C_u}{C_u + C_o}\right), \quad \text{number of historical forecast data points}\right). \quad (7.7)$$

The ratio $C_u/(C_u + C_o)$ is called the critical ratio, and it denotes the (optimal) probability of meeting all the demand for remanufactured products in the same period, given an underage cost C_u and an overage cost C_o. The number of data points in our case is 24, for 24 periods (months) of records of actual and forecasted demand (Table 7.1). For illustration, for product K5R, $C_u = 1000$ and $C_o = 1$, which indicates that backlogging demand for this product is not a good idea (i.e., the critical ratio is very high at 0.999). Applying (7.7), we obtain $k = \text{TINV}(2 * [1 - 0.00099], 24) = 3.467$. As a result, the SS multiplier is computed from (7.6) as SS multiplier = $1 + 3.467 *$ $0.1024 = 1.355$. Thus, the firm should add (fluctuating) SSs of remanufactured products in each period equal to 35 percent of the expected demand for that period.

We now show how we can translate these SSs into a familiar MRP table. Assume that demand forecasts (F) for remanufactured K5R products for the months of October 2008, November 2008, December 2008, January 2009, February 2009, and March 2009 are equal to 225, 250, 325, 335, 380, and 475, respectively. Assume the remanufacturing lead time for product K5R is equal to one month. Thus, the firm remanufactures in September to meet demand in October. The MRP table for the product K5R is displayed in Table 7.2, where each row is interpreted as follows.

The row "current period" reminds us that we are at the end of the period September 2008 (period 0). The row "forecast" is taken directly from remanufactured product K5R's demand forecasts (from Table 7.1 for periods May 2008–September 2008; future forecasts for periods October 2008 and beyond). The row "forecast * SS mult." is given by the row "forecast" multiplied by the SS multiplier (1.355). For example, for the month of May 2008, the entry in this row is 385 * 1.355 = 522. Thus, the target beginning inventory of product K5R for the month of May 2008 is 522 units. The row "actual" corresponds to actual demand for remanufactured products at each period. So, for example, for the month of May 2008, forecasted demand for product K5R was 385 units, but actual demand turned out to be 326 units. "On-hand actual" describes the on-hand inventory of product K5R at the end of the period, which is equal to 196 units (522 target inventory minus 326 actual demand) in May 2008. "On-hand projected" describes the expected inventory at the end of the period; this row is only relevant for future periods (i.e., periods October 2008 and beyond), and is always equal to the target inventory (forecast * SS multiplier) minus forecast.

Note that the firm has 155 product K5R units at the end of the period April 2008 (i.e., start of the period May 2008). Given a target inventory of 522 units and a starting inventory of 155 units, the firm has a "net requirement" (see corresponding row) of 367 units (=522 – 155) in the period May 2008. The row "planned order receipt" is then equal to the "net requirement" row, assuming the firm wants to meet the target inventory level. Finally, the row "planned order release" means the

Table 7.2 MRP Table for Product K5R

Month		May 2008	June 2008	July 2008	August 2008	September 2008	October 2008	November 2008	December 2008	January 2009	February 2009	March 2009
Current period		-4	-3	-2	-1	0	1	2	3	4	5	6
Forecast		385	495	360	215	310	225	250	325	335	380	475
Forecast * SS multiplier		522	671	488	291	421	305	339	441	454	515	644
Actual		326	530	354	187	288						
On-hand actual	155	196	141	134	105	133						
On-hand projected						111	80	89	116	119	135	169
Net requirement		367	475	347	158	316	195	259	352	338	396	509
Planned order receipt		367	475	347	158	316	195	259	352	338	396	509
Planned order release	367	475	347	158	316	195	259	352	338	396	509	367

Note: SS multiplier = 1.355; lead time: one month; current period: end of September 2008. Dark grey indicates present period and light grey indicates future period.

firm has to release a production order of 367 units at the beginning of the month of April 2008, given the one-month lead time for remanufacturing, to make sure that the 367 units are available to meet the demand at the beginning of the period May 2008.

Now, let us turn our attention to the remanufacturing plan going forward, that is, at the end of the period September 2008. Given the one-month lead time, the number of units remanufactured in the month of October 2008 is based on the demand forecasts for the month of November 2008. Demand forecast for the month of November 2008 is 250 units; thus forecast * SS multiplier = 250 * 1.355 = 339 units; this is the "target" inventory at the beginning of the month of November 2008. Given the projected inventory at the end of the month of October 2008 of 80 units (target of 225 * 1.355 – forecast of 225), the firm has a net requirement of 259 units (339 – 80) for the month of November 2008, which is precisely the planned remanufacturing production for the month of October 2008.

The next step for the firm is to plan the number of incoming returns (cores) to salvage, which is shown in Table 7.3. In our example, the number of returns at this firm has historically been much higher than the demand for remanufactured products. Specifically, the number of returns (from end of lease) received in the months of June 2008, July 2008, and August 2008 were 1167, 1069, and 1289, respectively, yielding a three-month average of 1175 returns. Given a requirement of 259 units to remanufacture in the month of October 2008, it is clear that the firm will remanufacture an expected fraction 259/1175, or 22 percent of all returns received in the month of September 2008. Thus, the firm should remanufacture, in the month of October 2008, the top 22 percent (in terms of quality) of the returns received in September 2008. If there are four quality grades, for example, and they arrive in equal proportions (25 percent of each type), then the firm should only remanufacture returns in the top-quality grade, salvaging the rest. If, on the other

Table 7.3 Planning the Number of Incoming Returns in September 2008 to Salvage and Remanufacture

Month	Returns Received
June 2008	1167
July 2008	1069
August 2008	1289
Three-month average	1175
Planned order release (October 2008)	259

Note: Expected fraction of returns to be remanufactured: 0.22 (259/1175). Thus, the firm should remanufacture the top 22 percent quality returns received in September 2008.

hand, the number of returns is about the same as the demand for remanufactured products, then the firm will need to store any returns not used for remanufacturing in the current period for use in future periods.

7.4 Conclusion

In this chapter, we provided two methodologies for production planning in remanufacturing. The first model is an optimization model that requires data such as the future forecasts for the number of returns of each quality grade, the demand for remanufactured products over the planning horizon, and the relevant costs (remanufacturing, salvaging revenue, holding cost, and backlogging cost). The second model is a heuristic approach that incorporates demand forecast uncertainty, a rolling planning horizon, and limited information on return forecasts and their quality levels. It uses an MRP-based logic, and can be easily implemented on a spreadsheet. While the first model has the advantage that it explicitly considers capacity constraints, the second model may be more appropriate when there is significant uncertainty in the demand for the remanufactured product.

References

Ferguson, ME, Guide Jr., VDR, Koca, E, and Souza, GC. The value of quality grading in remanufacturing. *Production and Operations Management* 2009; 18: 300–314.

Guide Jr., VDR. Production planning and control for remanufacturing: Industry practice and research needs. *Journal of Operations Management* 2000; 18: 467–483.

Souza, G. Closed-loop supply chains with remanufacturing. In Z. L. Chen and R. Raghavan (Eds.), *Tutorials in Operations Research*, pp. 130–153, 2008. Informs, Hanover, MD (available at www.informs.org).

Chapter 8

Market for Remanufactured Products: Empirical Findings

Ravi Subramanian

Contents

8.1 Introduction

Market acceptance of remanufactured products is key to the success of remanufacturing operations, notwithstanding the production cost advantage of remanufacturing as compared to manufacturing new products. Several factors affect the acceptance of remanufactured products, including price differences between equivalent new and remanufactured products, seller reputation, the nature of warranties offered, the level of purchasing experience among potential buyers, and customer perceptions about remanufactured products. In this chapter, we shed light on such demand-side factors and present empirical findings from our related research.

Although there are certain similarities between used products and remanufactured products, a very important difference is that used products are typically sold in an "as-is" condition whereas remanufactured products go through a variety of formal processes to ensure that the product performs just as good or better as its new counterpart. This difference also explains the marketing channels in which used and remanufactured products are typically transacted. Used products are typically transacted via classifieds, private exchanges, and auction sites (such as eBay®, mostly by individuals or bulk traders), while remanufactured products are typically sold via marketing channels for new products (such as OEM Web sites where both new and remanufactured products can be purchased),* as well as auction sites such as eBay, where large numbers of remanufactured products are sold by OEMs or their authorized distributors.

The markets for used and remanufactured products are similar in that both markets involve buyer uncertainty about product quality, resulting in lower consumer willingness to pay (WTP) than that for new products. However, remanufactured products may face higher levels of uncertainty among buyers as compared to used products, due to the nontrivial effort involved in bringing returning product cores to their original or updated specifications. In the presence of such uncertainty, the seller reputation and strong product warranties are critical to the market acceptance of remanufactured products and for sellers to command higher prices. Additionally, potential buyers who have limited knowledge of the formal processes involved in remanufacturing, or about the comparable performance of remanufactured and new products, may be less inclined toward purchasing remanufactured products. On the other hand, just as used products may attract customers who either would not buy or are unable to afford new products, remanufactured products may enable market expansion into customer segments that would not buy the more expensive, new counterparts.

Online auction sites—particularly eBay—have provided almost universal access to feedback on purchased remanufactured products and to reputation measures of sellers of such products. Apart from auction sites, the proliferation of online forums where participants voluntarily exchange their experiences with and concerns related

---

* Used products are occasionally sold alongside new products (e.g., used automobiles).

to remanufactured products has made it even more imperative that sellers of remanufactured products strive to build and maintain their reputation, communicate
the value proposition of remanufactured products, or otherwise exercise efforts to
alleviate buyer concerns regarding remanufactured product quality.

This chapter is organized as follows. In Section 8.2, we introduce our empirical study (Subramanian and Subramanyam 2008) on which this chapter is based.
In subsequent sections, we provide a discussion of our empirical findings related
to the price differentials between corresponding new and remanufactured products
(Section 8.3), the impacts of seller reputation and warranties on these price differentials (Section 8.4), buyer experience across new and remanufactured products
(Section 8.5), and buyer satisfaction with remanufactured products compared to new
products (Section 8.6). Section 8.7 concludes this chapter with managerial insights.

8.2 Empirical Study

As a major marketplace for remanufactured products and a rich source of information
on sellers, buyers, and buyer feedback on purchased products (Gomes 2004), eBay
serves as an ideal site for obtaining empirical data pertaining to the market factors
identified here as impacting the viability of remanufacturing. Significant volumes of
remanufactured products are sold on eBay by OEMs or their authorized distributors,
and by third-party remanufacturers or resellers. Our empirical study (Subramanian
and Subramanyam 2008) utilized detailed data from the years 2006 and 2007 on
completed purchases of remanufactured products on eBay, including prices paid by
buyers, seller and buyer reputation scores (also proxies for experience), and transaction-related feedback. The keywords used in our study to identify transactions of
remanufactured products were those recognized in the remanufacturing literature
(Hauser and Lund 2003), namely, remanufactured, refurbished, and rebuilt.

The first dataset in our study was compiled to assess the differences between
the prices of corresponding new and remanufactured products, and the impacts
of seller reputation and remanufactured product warranties on these price differences. WTP has been recognized both in practice and in the academic literature
to be challenging to measure. Market data (i.e., data on completed purchases)
has the advantage of representing actual purchase behavior; although purchase
prices only provide lower bounds on WTP, higher purchase prices at least cor-
relate with higher WTP. While the delivered prices (price paid plus shipping cost)
of remanufactured products were obtained from eBay, the reference prices for
simultaneously available, corresponding new products were obtained using price
search engines. The price differential between a new product and its remanufactured version was measured as the difference between the lowest delivered price
for the identical new product (obtained using price search engines) and the delivered price paid by the buyer of the remanufactured product on eBay, as a percentage of the delivered price of the new product. For each of the eBay transactions

of remanufactured products, item-related, seller-related, and buyer-related data from eBay were simultaneously extracted. The remanufactured products in our study were identified as originating from a variety of sources—including inspection failures, warranty and end-of-use returns, floor samples, and units damaged in shipping. Warranties for the remanufactured products were classified as OEM or authorized factory, third party, and none.

A second dataset of completed eBay transactions of both remanufactured and new products was assembled to empirically contrast buyer experience and feedback for remanufactured and new products. In addition to item-, seller-, and buyer-related data, feedback information for the seller and the buyer was also collected for each transaction, that is, whether positive, neutral, negative, or no feedback for the seller and buyer as well as feedback comments, if any. Additionally, content analysis of feedback comments in the sample was performed to isolate product-related impressions from transaction-related impressions. For example, the seller may have received positive feedback even if the product was defective but the seller offered a positive transaction experience (e.g., by allowing the buyer to return the defective product). The two datasets included transactions spanning the following eBay product categories: business and industrial, cameras and photo, cell phones and PDAs, computers and networking, consumer electronics, home and garden, jewelry and watches, musical instruments, and video games.

8.3 Price Differentials

> If the price is right, it makes sense to buy [remanufactured] equipment.
> I've bought a lot ... with no problems.
> *— from an online forum discussing remanufactured computers*

Buyer uncertainty about remanufactured product quality or even a possible lack of knowledge about what is entailed in remanufacturing would translate into lower WTP for remanufactured products as compared to new products. As remanufacturing a product is typically less expensive than manufacturing a new one and because of buyer uncertainty, it makes economic sense for remanufactured products to be offered at relatively lower prices compared to new products, irrespective of how successfully the comparable performance of remanufactured and new products is communicated to potential buyers.

Perhaps the first empirical finding pertaining to the lower WTP for remanufactured products was by Guide and Li (2007), who administered eBay auctions for new and remanufactured versions of two types of products (a consumer product and a commercial product). They found that the remanufactured versions of the products were purchased at lower prices than the new versions. Our study expanded the scope of product types considered in Guide and Li (2007) to the entire breadth of product categories in which remanufactured products are sold on eBay, with the

Table 8.1 Price Differential by Product Category

Product Category	Average Price Differential[a]	Transaction Volume[b]
Business and industrial	Mid	Mid
Cameras and photo	Mid	Low
Cell phones and PDAs	Mid	Low
Computers and networking	High	High
Consumer electronics	Mid	Mid
Home and garden	High	Low
Musical instruments	Low	Low
Video games	High	Low

[a] Low: 0–10 percent; mid: 10–30 percent; high: >30 percent.
[b] Low: 0–5 percent of the total number of transactions across all product categories; mid: 5–20 percent; high: >20 percent. The jewelry and watches category was excluded from the statistical analysis due to an insufficient number of transactions.

objective of finding whether, and how, price differentials vary by product category. Table 8.1 summarizes the average price differentials across the various product categories in our study.

Unique category-related aspects such as transaction volumes and rates of technological obsolescence play important roles in explaining the prices at which remanufactured products are purchased. For example, larger category-wise transaction volumes were generally found to be associated with higher price differentials in our study. Also, potentially because remanufacturable cores of relatively recent product generations as well as those of relatively older product generations are typically scarce, price differentials for models belonging to intermediate generations were found to be greater than the price differentials for products belonging to relatively older or newer generations.

Within our dataset, iPods® were identified as suitable candidates for exploring how price differentials vary with product generation when substitutable new and remanufactured products across multiple generations are simultaneously available. Our dataset included transactions of various remanufactured iPod models, for which we obtained product release dates from Apple's press releases. Figure 8.1 shows a plot of the average price differential with respect to the time of product introduction for the iPod models in our dataset, indicating the non-monotonic behavior described above. Apart from the product diffusion argument made above, another possible reason for such a behavior of price differential could be that prices

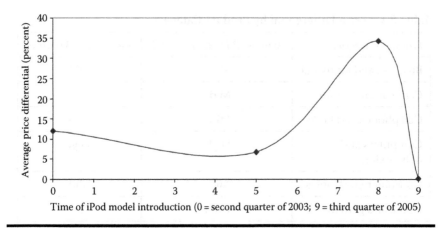

Figure 8.1 Price differential versus product generation (iPods). (Our analysis only included iPods released after 2003 because remanufactured iPods of generations earlier than 2003 did not have new counterparts simultaneously available at the time of our study.)

of new products of earlier model vintages are likely to decrease with time, and that new products of recent generations are likely to be in high demand.

8.4 Seller Reputation and Warranties

> I agree with the reputable seller part as well. I have been buying Dell refurbs for over 10 years and have received better service from them than with new machines. They have the same warranty and are much cheaper ... Laptops or desktops, I get great service from both. Try the Dell Outlet!
>
> *— from an online forum discussing remanufactured computers*

> I bought a refurbished Kodak from eBay. The flash stopped working about a month later. I got it repaired through Kodak using the warranty. I would not have made the purchase without the warranty and was very glad that I did. Other than that, the camera is excellent. So I vote yes to refurbished cameras WITH a valid warranty that includes enough time to be used, if needed.
>
> *— from an online forum discussing remanufactured cameras*

Under uncertainty about product or service quality, reputable sellers are likely to enjoy higher prices from credible communication of such reputation (Akerlof 1970). Reputation systems, such as that on eBay and Amazon.com®, provide commonly available and relatively objective information to potential buyers as to the trustworthiness of sellers. Feedback (positive or negative) left by previous transactors can be used as measures of reputation. In the literature, results pertaining to the effect of positive or negative seller reputation on purchase prices of used products (relative to reference prices) are mixed in that certain studies have been able to show the existence of an effect whereas other studies have not (Resnick et al. 2006).

As mentioned earlier, an important difference between used and remanufactured products is that buyer uncertainty may be more of an issue with remanufactured products due to the typically significant effort involved in testing and bringing products to their original or upgraded specifications. Consistent with our expectation, we found significant effects of both negative as well as positive seller reputation on the differentials between reference prices of corresponding new products and purchase prices of remanufactured products. Greater negative seller reputation is significantly associated with larger price differentials while greater positive reputation is significantly associated with lower price differentials. Thus, sellers with poor reputation have to offer significantly lower prices to attract buyers.

Additionally, for remanufactured products, seller reputation is a more important purchase criterion than for new products. The average negative reputation score for successful sellers of remanufactured products in our study was about one-third of that for successful sellers of new products. Overall, our findings highlight the importance of seller reputation in determining the prices that buyers are willing to pay for remanufactured products. Our study also found a significant effect of warranty strength for the remanufactured product on price differentials. Greater the strength of the warranty for the remanufactured product, the lower the price differential. Thus, strong warranties are also important for remanufactured products to command high prices.

8.5 Buyer Experience

> I had some reservations about buying a refurbished Kindle, but after reading the posts, I used the recommended link ..., and I guess I was just lucky, because there was 1 refurbished Kindle available, which I ordered ASAP. Interesting enough, there was a tab that reflected that there was 1 new Kindle available too, but since this is still a first generation device, I decided to save some $ and go with the refurbished to evaluate the "new" technology.
>
> *– from an online forum discussing remanufactured Kindles™ from*
> *Amazon.com*

My wife and I have had refurbished iPods: 2 G 10 GB, 2 G 20 GB, 3 G 20 GB, & 3 G 40 GB. No problems whatsoever except for a faulty firewire cord ... and they had a new one to me the next morning.

Everything is included except the original box and you still get a warranty. They are a great deal!
        — from an online forum discussing remanufactured iPods

Experienced buyers are more likely to make unbiased purchasing decisions by relying on objective indicators of product quality. In the context of remanufactured products, the lack of knowledge about the formal processes involved in remanufacturing would result in less-experienced buyers being less inclined toward remanufactured products. In their survey, Hauser and Lund (2003) found that expert buyers are less likely to be dissuaded simply because a product is labeled as "remanufactured."

Using the total of positive and negative reputation scores for the buyer as a proxy for experience, the second dataset in our study examined whether buyers of remanufactured products are, on average, more experienced than buyers of new products. In addition, we also explored how buyer experience for remanufactured products varies by product category. For example, buyers in a technologically more complex product category (such as computers and networking) can be expected to be more experienced, on average, than buyers in a less-complex category (such as cell phones and PDAs).

Interestingly, our study found that although certain categories (see Table 8.2) showed higher buyer experience for remanufactured products than for new products, the relationship was the opposite for other categories. In other words, for some categories, buyer experience was lower for remanufactured products than for new products. A probable reason for such a finding is market expansion due to the extended set of options available to potential buyers (Ghose et al. 2005). Also, sellers of remanufactured products typically invest greater levels of effort in communicating the processes involved in remanufacturing and the economic benefit to customers. Such efforts are more likely to induce purchases from less-experienced buyers as compared to more experienced buyers.

Table 8.2 also presents evidence of variations in buyer experience across product categories for remanufactured products. Our study found the following: (1) remanufactured products in the cell phones and PDAs and video games categories are purchased by relatively less-experienced buyers as compared to remanufactured products in other categories. (2) Remanufactured products in the jewelry and watches, musical instruments, and computers and networking categories are purchased by relatively more experienced buyers as compared to remanufactured products in other categories.

8.6 Post-Purchase Buyer Feedback

I've bought a lot of computers from several vendors with my own money over the years, and here's what I've found: New computers are sold with factory-level quality control ... When a company takes a computer back

Table 8.2 Median Buyer Experience by Product Category (Remanufactured versus New)

Product Category	Median Buyer Experience[a]	
	Remanufactured	New
Business and industrial	**Mid**	**High**
Cameras and photo	**High**	**Mid**
Cell phones and PDAs	Low	Low
Computers and networking	High	High
Consumer electronics	Mid	Mid
Home and garden	**Mid**	**High**
Jewelry and watches	High	High
Musical instruments	High	Mid
Video games	**Low**	**Mid**

[a] Low: <25; mid: 25–50; high: >50. A bold entry indicates a statistically significant difference between the median buyer experience across remanufactured and new products in the specific product category.

for any reason and wants to resell it, it is in their best interest to make sure it doesn't come back again, so [the] refurb process usually involves more stringent testing than the computer already received, AFTER the known flaw is fixed. This is a good thing. Chances of getting a bad one after all this testing are low.

— from an online forum discussing remanufactured Dell™ computers

Once the buyer receives the purchased remanufactured product and has the opportunity to experience it, the performance of the product assumes importance over concerns about seller reputation. Consumer satisfaction with a product is closely linked to prior expectations and the price paid (Anderson et al. 1994, Fornell et al. 1996). Remanufactured products may be associated with lower (prior) expectations relative to new products. Therefore, upon using remanufactured products, if buyers observe their performance as being equivalent to the performance of corresponding new products, satisfaction levels would indeed be superior for remanufactured products.

By measuring post-purchase satisfaction as whether product-related feedback in a transaction is positive or not (and accounting for product, reputation, and market-specific factors, such as purchase prices and potential for feedback retaliation*), our study found that remanufactured products are associated with a significantly higher likelihood of buyer satisfaction than new products. This finding can be attributed to both the observed "comparable to new" performance of remanufactured products and the possibly lower expectations associated with remanufactured products. Table 8.3 includes examples of positive feedback comments left by buyers of remanufactured products, reflecting the performance and price advantage of the remanufactured products purchased. Also, the significant difference between the likelihood of positive feedback for remanufactured and new products could also be due to the fact that remanufactured products are likely to be purchased from more reputable sellers in the first place, as compared to new products.

Table 8.3   Examples of Positive Feedback Comments for Remanufactured Products

"LCD TV works like its Brand New!!!!!! Fast Shipping A+++++++ Will Buy From Again"
"Great eBayer mower works like new … very happy thanks so much"
"Thanks for the motor! It works beautifully! My lux G is like new!!"
"Laptop works great. Great service. Highly recommend this seller. A+++"
"This seller saved me a ton of money off the retail price of my new TomTom, A+!"
"Words cannot describe the beauty of this sax … awesome … great ebayer … AAAAAAAAA"
"Great value … Remanufactured cartridge worked perfectly A+"
"The camera shipped quick, was in factory sealed box and works great."
"Great motherboards. All 6 work perfect!"
"PERFECT DRILL RIGHT PRICE"
"Printer Works GREAT! Excellent Value. Will Buy From Again. Thanks."
"If you need a papercutter this is the one! awesome quality! awesome price!"
"Shipping fast, PSP works as expected, and looks like new"
"NAIL GUN LOOKS NEW VERY SATISFIED"

---

* For example, a buyer who delays payment might leave negative feedback for the seller in response to justifiably receiving negative feedback from the seller.

8.7 Conclusion

One of the key challenges in the context of remanufacturing is how to set the price of the remanufactured product at a point that will just trigger purchases from prospective buyers. Pricing strategies employed in practice tend to be rather simplistic, for example, cost plus, or a certain percent below the price of new, rather than based on a formal assessment of buyer WTP. Although there are challenges involved in ascertaining WTP, there exist methods, such as those in our empirical study or in the experimental approach by Guide and Li (2007), that can provide insights into actual purchase behavior. For example, our study found that the prices (relative to the reference prices of corresponding new products) at which buyers purchase remanufactured products varies significantly by product category.

Seller reputation is extremely important for remanufactured products. Buyers of remanufactured products are highly sensitive to seller reputation, and sellers with poor reputation have to offer considerably lower prices to attract buyers. Therefore, in markets for remanufactured products, mechanisms that enable the communication of reputation are important for sellers of remanufactured products. Apart from seller reputation, the purchase prices of remanufactured products are significantly influenced by the strength of warranties offered.

In contrast to the intuitive expectation that the expertise of buyers of remanufactured products would likely be greater than that of buyers of new products (Hauser and Lund 2003), our study found that although there exist product categories for which buyers of remanufactured products are indeed more experienced than buyers of new products, the same is not true for other categories. One reason could be the possibility of market expansion to buyers who are unable to buy higher-priced new products. Variations in buyer experience across remanufactured product categories suggest that sellers should consider designing their marketing strategies in a manner consistent with characteristics of the target market. Offering generous return policies or stronger warranties may attract those buyers who are on the fence, debating between whether to buy a new or a remanufactured product.

Notably, remanufactured products are associated with significantly greater likelihood of positive feedback compared to new products. In other words, despite uncertainty in product quality prior to purchase, buyers of remanufactured products indeed perceive greater value in remanufactured products most likely due to comparable qualities, lower prices, and perhaps lower expectations.

The discussion in this chapter hints at opportunities for businesses to extract greater value from remanufactured products relative to current practice. Online marketplaces are ideal for attracting buyers of different demographics (location, expertise, etc.). Such marketplaces also provide opportunities for the seller to not only effectively communicate reputation but also provide additional information regarding the remanufactured product offering, including responding to buyer questions about the offering and making these responses visible to all prospective buyers.

A caveat to the findings discussed in this chapter is that our study focused on a particular online marketplace where buyers are unable to physically observe products prior to the purchase decision. The findings will have to be validated in other marketplaces—both virtual and physical. Additionally, complementary research methods such as surveys and experiments may be used to build upon or further validate our findings.

In concluding, we recognize that the sheer variety of remanufactured products that are currently being purchased by diverse types of buyers across various categories of products and technologies is in sharp contrast to the traditional view that only products belonging to slow-moving technologies are amenable to remanufacturing. Given the current gaps in both practice and theory, there is definitely a need for both practitioners and academicians to gain a deeper understanding of the market factors that drive the success of remanufacturing efforts.

References

Akerlof, G. A. 1970. The market for "Lemons": Quality uncertainty and the market mechanism. *Quarterly Journal of Economics* **84**(3) 488–500.

Anderson, E. W., C. Fornell, and D. R. Lehmann. 1994. Customer satisfaction, market share, and profitability: Findings from Sweden. *Journal of Marketing* **58** 53–66.

Fornell, C., M. D. Johnson, E. W. Anderson, J. Cha, and B.E. Bryant. 1996. The American customer satisfaction index: Nature, purpose and findings. *Journal of Marketing* **60** 7–18.

Ghose, A., R. Telang, and R. Krishnan. 2005. Effect of electronic secondary markets on the supply chain. *Journal of Management Information Systems* **22**(2) 91–120.

Gomes, L. 2004. E-commerce sites make great laboratory for today's economists. *The Wall Street Journal* (October 11).

Guide Jr., V. D. R. and K. Li. 2007. The Potential for Cannibalization of New Product Sales by Remanufactured Products. *Working Paper*. Pennsylvania State University, Pennsylvania, PA.

Hauser, W. and R. T. Lund. 2003. *The Remanufacturing Industry: Anatomy of a Giant.* Department of Manufacturing Engineering, Boston University, Boston, MA.

Resnick, P., R. Zeckhauser, J. Swanson, and K. Lockwood. 2006. The value of reputation on eBay: A controlled experiment. *Experimental Economics* **9** 79–101.

Subramanian, R. and R. Subramanyam. 2008. Key Drivers in the Market for Remanufactured Products: Empirical Evidence from eBay. Available at SSRN: http://ssrn.com/abstract=1320719.

INDUSTRY CHARACTERISTICS AND CASE STUDIES

III

INDUSTRY CHARACTERISTICS AND CASE STUDIES

Chapter 9

Examples of Existing Profitable Practices in Product Take-Back and Recovery

Mark Ferguson, Gilvan C. Souza, and L. Beril Toktay

Contents

9.1 Introduction

In practice, we observe that some firms of durable or semidurable goods remanufacture their end-of-use products while others choose not to. Insofar as remanufacturing requires different competencies than manufacturing, whether to enter remanufacturing is a similar question as whether to enter any new business. At the same time, this decision has characteristics of a product line extension decision, as the remanufactured product has the same functionality as the corresponding

145

new product. These strategic decisions, regarding whether a firm should or should not remanufacture, are discussed in Chapter 2. There are also certain characteristics of the remanufacturing business that make it unique (as opposed to a normal product line extension), and that are best appreciated by taking a process-based perspective.

In this chapter, we develop a framework that combines the strategic perspective with the process perspective, to analyze five industries where there is profitable remanufacturing activity. Our set of industries is not meant to be comprehensive, but rather to represent a wide range of industries with distinct environments and market dynamics. The selected set includes industries where the original equipment manufacturers (OEMs) are actively involved in remanufacturing and reselling their used products as well as industries where the OEMs are rarely involved and the bulk of remanufacturing is performed by third-party firms. From our observations across this industry set, we propose some common characteristics of an industry along each dimension that may make a firm more or less likely to find it profitable to be actively involved in remanufacturing.

Our discussion takes a profit-based perspective, following the majority of the operations management literature. At the same time, there are significant environmental implications of a firm's decision to remanufacture or not that are increasingly relevant to businesses today; however, this is beyond the scope of this chapter.

9.2 When Is Remanufacturing Attractive? Observations Based on Practice

We build a framework that describes characteristics of an industry that have a potential impact on remanufacturing attractiveness, including a more granular, process and technology perspective that complements the strategic analysis of Chapter 2. Hauser and Lund (2003) propose a set of six "remanufacturing-friendly" conditions:

1. *Nature of the product*: The product should have a core—a remanufacturable entity. It should have a long useful life, and be based on stable technology. Further, a large portion of the product's original cost should be value-added (labor, energy, capital) compared to purely material costs.
2. *Type and size of the market*: Buyers of remanufactured products should have significant expertise in evaluating products, so industrial markets are prime targets for remanufactured products.
3. *Availability of returns*: The remanufacturer should have the ability to recover a significant number of returns at a reasonably low cost.
4. *Supporting infrastructure*: There should be a distribution network and transportation system for supplies needed for remanufacturing as well as finished remanufactured products.

5. *Remanufacturing technology*: There should be trade associations, publications, and even universities that provide remanufacturers with the technology and know-how they need.
6. *Legal and regulatory leeway*: The regulatory environment should not prohibit, excessively tax, or establish unreasonable risks to the remanufacturer (in terms of product liability, intellectual property, and warranty law).

The framework by Hauser and Lund (2003) is helpful in providing guidance to a firm in its decision to remanufacture; however, it is based on a definition of remanufacturing as a thorough process that involves disassembly, value-added activities (cleaning, repairing, replacing parts in bad condition), and reassembly. Thus, the conditions above are slightly biased toward larger items with mechanical or movable parts. We adopt a broader definition of remanufacturing: we view it as a value-added process that transforms a product that has been in a consumer's possession for some time into a remarketable product again; this process should preserve a considerable portion of the external shape of the product. Thus, our definition also includes refurbishing (e.g., sometimes consisting of simple testing and cleaning, conducted on consumer returns), but does not include recycling (i.e., materials recovery) or disassembly for spare parts. Our broader definition is in line with the literature on closed-loop supply chains prevalent in the operations management community. Under our broader definition, remanufacturing presents unique characteristics, as we describe below, that may require different capabilities than regular manufacturing. Guide and Van Wassenhove (2003) provide a process view of remanufacturing and the reverse chain, and indicate that most reverse chains are comprised of five fundamental activities:

1. *Product acquisition*: The input to the remanufacturing process—product returns—is uncertain in quality, quantity, and the timing of arrival. The firm may need to provide incentives to customers or resellers to be able to recover a sufficient amount of returns in the right quality to meet its needs. The collection cost is a key driver of the overall profitability of a remanufacturing operation, as discussed in Chapter 6.
2. *Reverse logistics*: The network for collecting returns is different than the supply chain for procuring components for regular manufacturing. Typically, returns originate from many different locations, and thus they require consolidation centers that make transportation cost-effective. In a sense, this is similar to the role of warehouses in a forward supply chain, and thus a firm may decide to use the forward distribution network to collect returns from collection points (resellers, customers, etc.), and aggregate them at a consolidation center for shipment to remanufacturing facilities. The academic literature (see Chapter 5) indicates that this decision is not always optimal, and it may be better to maintain separate networks for forward and reverse chains.

3. *Testing and disposition*: Some returns are found to be unfit for remanufacturing, and they must be scrapped or recycled. Alternatively, the firm can also salvage returns for spare parts, or sell them to a broker, as discussed in Chapter 6.
4. *Remanufacturing*: Returns with different quality grades have different remanufacturing processing times, costs, and capacity usage.
5. *Remarketing*: Consumers generally view remanufactured products as imperfect substitutes for new products, so they require different pricing, positioning, and channels than new products, as discussed in Chapter 8. Thus, the firm needs additional remarketing capabilities.

When an OEM decides to remanufacture its own product, it has sufficient knowledge of the product such that choosing to offer remanufactured versions of its products could be considered a product line extension. However, as detailed above, remanufacturing requires different capabilities, which may not overlap well with the firm's forward supply chain. For example, Cisco has an almost monopolistic position in its core (new) IT networking equipment business; however, it lags significantly (in volume) in the refurbished equipment market. Two key order winners in the refurbished IT market are delivery lead time (typically less than 48 hours) and price. With an entirely outsourced supply chain, it becomes difficult for Cisco to be as responsive as smaller (mostly local) refurbishers. In fact, Cisco's remanufacturing volume is much smaller than its competition.

In fact, one may view entering remanufacturing as entering a new industry. Porter's five forces model is a useful framework for a firm that is considering entering a new industry. Specifically, it posits that a firm's profitability is impacted by industry characteristics:

1. *Bargaining power of suppliers*: This is related to supplier concentration and proprietary technology. For remanufacturing, the primary issue is the source of returns, which is the main input into the remanufacturing process.
2. *Bargaining power of customers*: This is related to switching costs by customers. For remanufacturing, the questions are the following: How do consumers use remanufactured products? Are they used as substitutes for new products? The answers to these questions are also related to the technology of the remanufactured product, that is, if remanufactured and new products belong to the same technological generation or not.
3. *Threat of new entrants*: This is related to barriers to entry in the industry. For remanufacturing, barriers to entry are for the most part low as remanufacturing is labor intensive (Hauser and Lund 2003), there are no laws preventing firms from remanufacturing used products (i.e., no intellectual capital infringement), and distribution channels are abundant, particularly on the Web.

4. *Threat of substitute products or services*: This is related to competition from outside the industry. In remanufacturing, this would include other (non-remanufactured) products that perform the same function. An example that is often mentioned in the industry is the competition posed by "cheap imports."
5. *Intensity of rivalry*: This is related to industry concentration and the number of competitors. Most remanufacturing industries are fragmented, with hundreds or thousands of small players (Hauser and Lund 2003), so competition is fierce.

Consistent with the strategic management literature, and the process view of closed-loop supply chains, the OEM should view the decision to enter remanufacturing as a function of "industry" attractiveness, existing resources, and operational capabilities. In the next section, we describe industry practice, where we choose description categories that are consistent with the frameworks above:

- Technology, use of remanufactured products (bargaining power of customers)
- Sources of returns (bargaining power of suppliers)
- Competitive landscape (number of competitors)
- Reasons for OEMs remanufacturing and extent to which top three OEMs remanufacture (intensity of rivalry, threat of substitute products and services)
- Barriers to entry (threat of new entrants)
- Collection cost, acquisition cost, sources of returns (product acquisition, reverse logistics)
- Salvage options (testing and disposition)
- Remanufacturing cost (remanufacturing, reverse logistics)
- Remanufacturing process (remanufacturing)
- Pricing, channels, market size (remarketing)

9.3 Remanufacturing Practice

Given the dimensions chosen to represent different remanufacturing industries (technology, sources or returns, competitive landscape, costs, pricing, and channels of distribution, etc.), it is clear that remanufacturing practice varies considerably across industries. This is highlighted in the survey by Hauser and Lund (2003), who collected data from 274 American remanufacturing firms in five industries—automotive, electrical equipment, furniture, machinery, and toner cartridges. We here provide an updated snapshot of practices in five representative industries: IT networking equipment, toner cartridges, PCs and printers, retreaded tires, and single-use cameras. We organize the discussion in five different tables. For

Table 9.1 Remanufacturing Practice: IT Networking

Dimension	Secondary IT Networking Industry
Technology	Typically, remanufactured and new products do not belong to the same generation, but they may (e.g., product returns) [interview with Canvas Systems (GA) senior manager, Steven Hyser] [0]
Use of remanufactured products	Network expansion, cheaper and quicker full replacement for new products, redundancy, spare parts, testing, and training [http://www.uneda.com/files/UNEDAMembershipSurveyResults.pdf] [+]
Pricing	Between 10 and 90 percent off a new product's list price (Sheldon 2007). An OEM (Cisco) offers remanufactured products at 25–30 percent off new [+ +]
Market size	According to UNEDA (United Network Equipment Dealer Association), $2 billion [www.uneda.com] [+ +]
Channels	Online channels; larger firms have dedicated sales agents [interview with Canvas Systems (GA) senior manager, Steven Hyser] [+]
Sources of returns	Technology upgrades, asset sales of firms who have gone out of business [0]
Acquisition cost	Remanufacturers offer "cash or trade" programs to firms wanting to sell their network equipment. Market prices exist, at least for the most popular equipment [visit to Canvas Systems 08/2007] [0]
Remanufacturing cost	5–20 percent of the cost of new equipment [visit to Cisco 02/2006 and Canvas Systems 08/2007] [+ +]
Competitive landscape	Cisco has a limited (outsourced) remanufacturing program, but there are many third-party providers—in excess of 300 firms [www.uneda.com] [–]
Reasons for OEM remanufacturing	According to Cisco, to provide more choices for customers, to sell more equipment, and to increase revenue from services (tech support) [visit to Cisco 02/2006] (+)
Service versus product model	Product, but with associated technical support services [–]
Extent to which top three OEMs remanufacture	Lucent (yes), Cisco (limited)

Table 9.1 (continued)   Remanufacturing Practice: IT Networking

Dimension	Secondary IT Networking Industry
Collection cost	Unclear
Salvage options	Remanufacture, sell as-is to brokers, recycle
Remanufacturing process	Disassembly of main modules, testing, and reconfiguration according to customers' specs [visit to Canvas Systems 08/2007] [+ +]
Regulatory context	Take-back legislation in the EU (WEEE), some states in the United States, Japan, China, and Taiwan [+ +]

each dimension, we also provide a score (e.g., [–]) indicating how that industry characteristic favors remanufacturing. For example, in the IT networking industry, the remanufacturing process is relatively simple, consisting mainly of disassembly, testing, and reconfiguration, thus favoring remanufacturing (score [+ +]), as the process does not necessitate significant investments.

Table 9.1 summarizes the secondary IT networking equipment industry. This is a large (over $2 billion a year) and highly fragmented industry, with over 300 third-party remanufacturers in the United States, along with OEMs such as Cisco and Lucent, although Cisco has a limited remanufacturing program. Typically, remanufactured products are one generation or more behind new products, and as a result, they are priced between 10 and 90 percent off the new product's price. Remanufacturing here is a low-cost operation (about 5–20 percent of the cost to manufacture a new product), consisting mainly of disassembly into modules, testing, the replacement of wearable parts, and reconfiguration. Used products are obtained from firms upgrading to newer technology equipment, firms going out of business, or product returns. Most remanufactured products are sold via online channels, although some firms have a dedicated sales force.

Table 9.2 summarizes the printer/copier cartridge refilling industry, which tends to be fought by the major OEMs and dominated by third-party refillers. These third-party firms compete for used cartridges, often offering free return shipping or trade-in rebates. The OEMs, who do not refill, also offer free shipping for empty cartridges but for the purpose of recycling them to keep them from being used for refilling. Refilled cartridges are often offered at significant discounts relative to new cartridges, but the third-party firms face a problem with the customers' perceptions of lower quality. These perceptions are driven by a combination of some disreputable refillers (due to the fragmentation of the industry) and by the actions of the OEMs such as the voiding of printer warranties when refilled cartridges are used. In response, the industry is beginning to consolidate and large office product retail chains are gaining market share with their own private label refilled

Table 9.2   Remanufacturing Practice: Toner Cartridges

Dimension	Toner Cartridges
Technology	Refilled toner cartridges must be of the same generation as new but the technology does not change very quickly [+]
Use of remanufactured products	Businesses and consumers who purchase printer/copiers from the OEMs but want to reduce their operating costs [+]
Pricing	Generally around 50 percent off of a new cartridge's list price [www.zdnet.com, www.inkreplacement.com] [+ +]
Market size	According to a 2005 report by Lyra Research, the refill market will grow to over 40 million annually by 2009, representing approximately 6 percent of the total cartridge market [lyra.ecnext.com/coms2/summary_0290–221_ITM] [+]
Channels	Online, mail, shopping center retail stores, business-to-business partnerships, and office supply superstores such as Office Depot and Staples [www.inkjetrefills.com] [+]
Sources of returns	Trade-ins and recovery via service calls. Some customers take their empty cartridge to a retail location to be refilled. The availability of used cartridges is a major constraint on this industry [0]
Acquisition cost	There is often no money paid for a returned cartridge but the return shipping is often paid for by the refiller [+]
Remanufacturing Cost	The cost of refilling a used cartridge is around 80 percent lower than making a new cartridge [www.inkjetrefills.com, www.cartridgeworldusa.com] [+ +]
Competitive landscape	Thousands of independent refillers but the major office supply superstores are gaining market share [www.rechargermag.com] [–]
Reasons for OEM remanufacturing	NA (0)
Service versus product model	Currently a combination as some refillers contract with local businesses for all of their cartridge needs [+]
Extent to which top three OEMs remanufacture	None

Table 9.2 (continued)   Remanufacturing Practice: Toner Cartridges

Dimension	Toner Cartridges
Collection cost	Typically depend on customers to return used cartridges
Salvage options	Recycling is common among the OEMs but is also costly
Remanufacturing process	Refilling a cartridge is relatively simple but refillers have to keep up with the latest technology developed by the OEMs to make refilling more difficult [+]
Regulatory context	Take-back legislation in the EU (WEEE), some states in the United States, Japan, China, Taiwan [+ +]

cartridges. So far, however, the OEMs continue to find it more profitable to fight the refilled cartridge market rather than to join it.

Table 9.3 summarizes the remanufactured computer and printer industry. Here, it is important to make a distinction between "newer" remanufactured PCs and printers (i.e., those from the same generation as new products) and "older" PCs and printers, as these are two fundamentally different industries. Newer remanufactured PCs and printers originate primarily from consumer returns within the grace period for returns, and they are offered primarily by the OEMs, who typically outsource the remanufacturing process itself of consumer returns to contract manufacturers, just like they do with new products. Due to on-par technology with new products, remanufactured PCs and printers are priced closer to new products, between 10 and 25 percent off the new product's price, typically with the same warranty. However, they are offered in separate channels—online and offline outlet stores—to avoid cannibalization. The remanufacturing process here is very limited, consisting primarily of minor cosmetic repairs, testing, and repackaging, consuming less than 10 percent of the cost of a new product. In contrast, "older" remanufactured PCs and printers originate primarily from end-of-lease returns, one or more generations behind (two to three years) new products. These products are remanufactured primarily by third parties and offered in online channels; they are priced at deep discounts (up to 80 percent off a new product's price), and they are typically consumed by nonprofits or consumers in developing countries.

Table 9.4 summarizes the remanufactured (retreaded) tire industry, which is significant only for trucks in the United States. Tire technology changes infrequently, and as a result retreaded and new tires belong to the same generation. Many commercial fleet companies have a service agreement with tire retailers, paying them on a mileage basis. As a result, tire retailers have an incentive to use lower-cost retreaded tires (than new tires) to replace nonfunctioning tires to maintain the vehicle in operation. The acquisition of returns is automatic upon a tire replacement. About half of all tires acquired by commercial fleet companies in the United States are retreaded. The retreading operation is comprehensive, consisting

Table 9.3 Remanufacturing Practice: PCs and Printers

Dimension	Newer PCs and Printers	Older PCs and Printers
Technology	Technology is similar to new products; origin is consumer returns [+ +]	Older technology, say from the end of lease (e.g., www.tigerdirect.com) [– –]
Use of remanufactured products	Near-perfect substitute for new products; frequently offer some warranty as new [0]	Technology can be used in nonprofit organizations, or emerging markets [–]
Pricing	Between 10 and 25 percent off a new product's list price [+]	Up to 80 percent off a new product's price [– –]
Market size	About 5 percent of sales of a new product [+]	For every two new PCs shipped to mature markets, one PC is resold in the secondary market (Gartner, Inc. 2005) [+]
Channels	Online and offline outlet stores by the OEM [0]	Online and offline outlet stores by third parties [0]
Sources of returns	Consumer returns up to 30 days after purchasing a new product [+ +]	End-of-lease returns (typically two to three years old) [– –]
Acquisition cost	Consumer returns are sent back to the OEM, who issues a full credit to the retailer [+]	Unclear
Remanufacturing cost	Low (<10 percent of cost of new product) [+ + +]	Similar to newer PCs and printers [+]
Competitive landscape	Only OEMs offer newer refurbished products as they are the only source of consumer returns [+]	Some OEMs refurbish older-technology PCs after the end of lease (e.g., IBM); thousands of third parties offer refurbished products of unknown origin and older technology [– – –]
Reasons for OEM remanufacturing	To increase profitability from recovering significant value in consumer returns [+ +]	To increase profitability [+]

Table 9.3 (continued) Remanufacturing Practice: PCs and Printers

Dimension	Newer PCs and Printers	Older PCs and Printers
Service versus product model	Product, but with associated technical support services [–]	Product [–]
Extent to which top three OEMs remanufacture	All major OEMs remanufacture consumer returns [+ +]	IBM [–]
Collection cost	Consumer returns are transported from retailers to manufacturers using contracts with 3PLs such as UPS [+ +]	Unclear
Salvage options	Remanufacturing, spare parts, use in service (warranty) [+]	Remanufacturing, recycling, spare parts [0]
Remanufacturing process	Mostly testing, cosmetic remanufacturing (replacing scratched plastic parts), replacing a single malfunctioning component, or software update [+ +]	Similar to newer-generation products [0]
Regulatory context	Take-back legislation in the EU (WEEE), some states in the United States, Japan, China, Taiwan [+ +]	Same [+ +]

Sources: Guide Jr., V.D. et al., *Manage. Sci.*, 52, 1200, 2006; visits to HP and IBM.

Table 9.4 Remanufacturing Practice: Retreaded Tires

Dimension	Retreaded Tires
Technology	Remanufactured and new products belong to the same generation as the tire technology does not change very often, especially for commercial tires (trucks, aircraft, heavy equipment) [+ +]
Use of remanufactured products	Retreaded tires are typically used as a full replacement for new products for commercial fleets. There are some laws against using retread tires on the steering axle. Commercial trucking companies will sign a service agreement with a tire store (retreader) that guarantees the trucking company "good" tires for a certain number of miles. The tire store may replace worn tires with new or retread tires [+ +]

(continued)

Table 9.4 (continued) Remanufacturing Practice: Retreaded Tires

Dimension	Retreaded Tires
Pricing	NA
Market size	Of the 37 million replacement tires purchased by commercial trucking fleets in 2004, nearly half were retreads [www.retread.org] [+ +]
Channels	Many independent tire retreaders who license retreading technology, buy retreading equipment, and receive training from one of the three main tire companies: Michelin, Bridgestone, and Goodyear [www.retread.org] [+ +]
Sources of returns	When a company under a service agreement has a tire replaced, the old tire is returned to the tire store
Acquisition cost	Hard to say because of the service contracts. The retreaders also buy some old tires occasionally to supplement the ones recovered through the service contract [+ +]
Remanufacturing cost	Twenty-five to thirty percent off the price of a new tire [http://www.retread.org/packet/index.cfm/ID/14.htm] [+ +]
Competitive landscape	Many small- to medium-sized retreaders that use one of the technologies developed by three large tire manufacturers. Within a particular geographic location, there are typically less than three competitors [+]
Reasons for OEM remanufacturing	To meet, in the most cost-effective manner, the demand for service contracts from large commercial trucking companies. Independent tire dealers initially created their own retreading processes but the OEMs developed higher-quality technology to protect their brand image. The OEMs also make money by licensing the technology and selling the rubber for the retreading process [personal interview with John Ziegler, VP of operations at Ziegler Tire] [+]
Service versus product model	Service model
Extent to which top three OEMs remanufacture	Top three OEMs offer remanufacturing technology that they license to third-party firms [www.moderntiredealer.com]
Collection cost	Minimal because of service contracts

Table 9.4 (continued)   Remanufacturing Practice: Retreaded Tires

Dimension	Retreaded Tires
Salvage options	Many uses for salvaged tires including burning for fuel or grinding and using recycled material for playgrounds and roads. In the United States, in 2007, 93 percent of all disposed tires are kept out of landfills [personal interview with Ralph Hulseman, Manager of Science and Expertise at Michelin] [+]
Remanufacturing process	Inspection of incoming cores using a special piece of equipment to check for nail holes or defects. If the casing is ok, then the remaining tread is scraped down to a standard diameter. A small layer of rubber strips are added to the casings and then the outer tread is rolled on. The tread is bought from the OEM and is proprietary to their process although it can be applied to any type of casing (e.g., another OEM's casing). A final baking attaches the tread [personal interview with John Ziegler, VP of operations at Ziegler Tire] [+]
Regulatory context	Forty-six states have banned the disposal of waste tires in landfills [www.p2sustainabilitylibrary.mil/P2_Opportunity_Handbook/7_I_A_10.html] [+ +]

of several steps, has been developed by the OEMs, and licensed to third-party firms. OEMs also profit from retreading by selling rubber for the retreading process. Tires can only be retreaded a finite number of times, and tires that are unfit for retreading are used for energy recovery or grinding for use in roads, as almost all states in the United States ban the landfilling of used tires.

Table 9.5 summarizes the single-use camera industry. Single-use cameras face a declining market share of the overall camera market due to the revolution brought about by digital cameras; however, they are still present. Single-use cameras are a natural product for remanufacturing, given the ease of collection (customers return used cameras to retailers for processing), and the significant value remaining in the product. The remanufacturing process is simple: disassembly, testing, and reassembly; and it saves OEMs about $2 per camera. In fact, large OEMs (Kodak, Fuji, Konica) reuse about 80 percent of recovered components from single-use cameras. As a result, most single-use cameras sold are likely to have one or many reused components, and therefore there is no price or physical difference between remanufactured and new products; they are sold in the same retail network.

What is clear from Tables 9.1 through 9.5 is that these industries have many features that make them attractive to remanufacturing, from strategic considerations

Table 9.5 Remanufacturing Practice: Single-Use Cameras

Dimension	Single-Use Cameras
Technology	The main change between product generations is size and appearance (smaller and less boxlike), two to three generations on market simultaneously, parts are not used between models although they may be the same due to the remanufacturing process, or between generations due to decreasing size (Muir 2006) [0]
Use of remanufactured products	One product only, may or may not contain reused interiors [+ +]
Pricing	N/A
Market size	22 M units sold in the United States in 1992, 150 M sold by Kodak since 1987 (Muir 2006) [0]
Channels	Sold through retail network [0]
Sources of returns	Customers take cameras to photoprocessors, who collect and ship them in bulk to Kodak for a rebate [+ +]
Acquisition cost	Rebate to photoprocessors, approx. $0.25 (Muir 2006) [+]
Remanufactured cost	Disassembly, testing, reassembly, estimated $1 per camera, material savings estimated $1 per regular and $3 per flash camera, or net savings of $2 per flash camera only (Muir 2006) [+]
Competitive landscape	Kodak has more than 70 percent of the U.S. market share. "Reloaders" illegally put film in the camera and sell as a brand camera, typically sourcing shells (e.g., Fuji) in Asia, as the manufacturers control the U.S. reverse flow well through channel relationships. At one time represented 10 percent of market [WSJ 2002] [+]
Reasons for OEM Remanufacturing	The goal was to enter new market segment of cost-conscious or impulse/event/convenience-based consumption with a low-cost disposable camera, remanufacturing program developed in response to environmentalist concerns
Service versus product model	Product
Extent to which top three OEMs remanufacture	Kodak, Fuji, Konica all can reuse/recycle 80+ percent by weight of returned cameras; returns estimated 50–80 percent for Kodak in the United States (Toktay et al. 2000, Muir 2006)

Table 9.5 (continued)   Remanufacturing Practice: Single-Use Cameras

Dimension	Single-Use Cameras
Collection cost	Parcel delivery from photoprocessors, estimated at ≈$0.30 at $2/lb shipping rate (Muir 2006)
Salvage options	Plastics recycling estimated to lose money (Muir 2006), used batteries donated, no other salvage options
Remanufacturing process	Simple manual partial disassembly (outer shell and lens), test, and assembly [+]
Regulatory context	Only for batteries

to more tactical, process-based considerations. We believe that our proposed dimensions can be used to analyze other industries to aid firms, especially OEMs, in their decision to remanufacture or not.

References

Guide Jr., V.D., G. Souza, L.N. Van Wassenhove, and J.D. Blackburn. 2006. Time value of commercial product returns. *Management Science* 52, 1200–1214.

Guide Jr., V.D. and L.N. Van Wassenhove (Eds.). 2003. *Business Aspects of Closed-Loop Supply Chains*. Pittsburgh, PA: Carnegie Mellon University Press.

Hauser, W. and R. Lund. 2003. *The Remanufacturing Industry: Anatomy of a Giant*. Boston, MA: Boston University.

Muir, M.C. 2006. Lifecycle assessment for strategic product design and management. MS thesis, Georgia Institute of Technology, Atlanta, GA.

Sheldon, M. 2007. Pre-owned gear gains foothold. *Communication News* 44(2). Available at www.comnews.com/features/2007_february/0207pre-owened.aspx.

Toktay, L.B., Wein, L.M., and S.A. Zenios. 2000. Inventory management of remanufacturable products. *Management Science* 46(11), 1412–1426.

Chapter 10

Reuse and Recycling in the Motion Picture Industry*

Charles J. Corbett

Contents

* This chapter is based on the chapters on "Environmental impacts of the motion picture industry: The micro view" and "Environmental best practices: Case studies" in Corbett, C.J. and R.P. Turco, *Sustainability in the Motion Picture Industry*, CIWMB Publication # 440-06-017, 2006. © 2006 by the California Integrated Waste Management Board. Used by permission.

10.1 Introduction

This chapter examines closed-loop supply chains in the context of the motion picture industry. It is based on the report resulting from the "Motion Picture Industry Sustainability" project conducted through the UCLA Institute of the Environment, under contract to the California Integrated Waste Management Board (CIWMB), during Summer 2003–Spring 2005.* The objectives of that study were to identify existing environmental best practices within the industry, focusing exclusively on production, not on distribution or on content.

CIWMB selected the motion picture and television industry due to its high visibility: environmental best practices uncovered while studying the motion picture industry are more likely to attract interest from and to be implemented by other industries. In addition, the motion picture industry can be thought of as a model toward which many other industries are converging. Industries such as fashion, toys, technology, aerospace, and pharmaceuticals all increasingly rely on a network of organizations jointly performing research and development, sharing responsibility for production and distribution, and disbanding after a few years when the next generation of products requires an entirely new network. This decentralized structure has been in place in the motion picture industry for many years, so understanding how the motion picture industry has dealt with environmental challenges is instructive for a broader set of industries.

This perspective applies equally to the value of examining closed-loop supply chains in the motion picture industry. While it is already challenging enough to design and manage closed-loop supply chains in well-established industries with stable supply chains, it becomes even harder to do so when the network of parties involved in a supply chain keeps changing, as is increasingly the case in industries such as those mentioned above. In this chapter, we discuss examples of closed-loop supply chains in the motion picture industry, including some more detailed case studies on the reuse of set materials.

For this study, we conducted semi-structured interviews with 43 individuals from a range of areas within the motion picture and television industry. We

---

* The motion picture and television industry has implemented a range of new initiatives since the completion of that study; those initiatives are not covered here.

covered most key functions above and below the line,* studio representatives and independents, from the business and creative sides of the industry, from the film and television side, and several from governmental and private organizations associated with the industry. The main missing categories are actors (who are notoriously hard to get access to) and directors of photography (who play an important role during the shoot).

10.2 Overview of the Industry

To put the examples of closed-loop supply chains in the motion picture industry, which we describe in the next section, in perspective, we first sketch the structure of the industry, starting with an overview of the players, then some comments on the operations of the industry, as reflected in the budget and the schedule.

10.2.1 The Players

One key feature of the industry is the degree to which it is decentralized. The seven major studios are the most visible part of the industry; most major distribution companies are owned by these seven major studios. Scott (2002, p. 962) provides an overview of the linkages between these seven studios and their subsidiaries.[†] However, a large part of the work is done by individuals who are temporarily employed by production companies for (part of) the duration of a specific film or television project. These production companies may be linked to any of the major studios, or may be independent. The independent production companies exist in a largely separate sphere from the major studios; according to Scott (2002,

---

* Singleton (1996, p. 8) explains the terms "above-the-line" and "below-the-line" as follows: "All feature budgets [...] retain the style of *the studio system*, in which a line is drawn across the top sheet. Above that line (Above-the-Line) are all the so-called artistic or creative components. Below (Below-the-Line) are all the technical and mechanical components." Singleton (1996, p. 413) elaborates: "Above-the-line expenditures are usually negotiated on a run of show basis and, generally, are the most expensive individual items on the budget. They include costs for story and screenplay, producer, director, and cast. Below-the-line costs include technicians, materials and labor. Labor costs are usually calculated on a daily basis. Also included in below-the-line costs are: raw stock, processing, equipment, stage space and all other production and post production costs."

† Scott (2002) lists the following seven majors: The Walt Disney Company (which includes Miramax, Buena Vista, Touchstone, and Dimension), Sony Corporation (including Columbia Tristar), AOL Time-Warner (including Warner Bros., Castle Rock, New Line, and others), Metro-Goldwyn-Mayer Inc. (including United Artists and Orion), News Corporation (including Twentieth Century Fox Film), Vivendi-Universal, and Viacom (including Paramount). Metro-Goldwyn-Mayer, Inc. recently merged with a consortium that includes Sony Corporation of America and Comcast Corporation. See http://www.mgm.com/corp_news_releases.do?id=424, last accessed April 25, 2005.

pp. 962–963) they "rarely come into contact with a major, and work in an entirely separate sphere of commercial and creative activity." That said, individuals may sometimes work for a production company affiliated with a major studio, and other times with independent production companies.

Storper (1989) describes how the vertically integrated studio system was gradually transformed into the vertically disintegrated system that prevails today. DeFillippi and Arthur (1998) describe the challenges faced by production companies as a result of growing from 0 to 150 or more employees within a few months, and then winding down to almost none a year or two later. These 150+ employees cover a very wide range of functions, including the obvious categories such as writers, actors, directors, and producers, and also wardrobe coordinators, animal handlers, generator operators, and many others.

A good overview of functions involved in motion pictures and television production, both above the line and below the line, is given in the U.S. Department of Commerce Report on Runaway Production (2001, p. 11). Jones (2003) breaks this down further, showing which personnel are typically needed in films of various budget sizes. The executive producer, the producer, and the director are the central figures in a production unit, where the balance of power varies between film, where the producer and especially the director are key, and television, where the executive producer is more influential. The producer hires the line producer, who in turn hires the department heads. "Each set has a different emotional tenor; different values and culture. But, for television, the network has ultimate decision-making power."*

Television is run more like a business than film; one interviewee characterized the bulk of film as "rich person's play." Indeed, one interviewee believed that "television would be a better part of the industry to focus on in implementing environmental practices, as it's easier to control." By contrast, the realm of commercials is even more dispersed among small companies that may have less environmental practices in place than any other part of the industry.

In one interviewee's words: "The producer and line-producer are the ones who have to create a system and set the tone for an environmentally conscious production." Some producers do strongly care about the environmental impacts of their work; a costume designer noted the example of a producer who "is always good about making others aware of the environment." At the first major production meeting he will say to us, "Let's make sure we're aware of our location, that we respect it and leave it the way we found it." The change in behavior has to be driven by the top down. However, the producer does not always have much direct influence over the crew; sometimes the producer is respected at a distance, sometimes not, though they do have more influence in a studio production. The line producer is the person who will mandate any specific practices, such as telling craft services to recycle.

---

* All quotes without sources are from the interviews conducted during the original study.

The director hires the art director and the director of photography, while the producer has little to say. The producer hires people to come and clean up after the shoot. The director of photography is the "below-the-line king," generally well educated, and more likely to be environmentally aware, as are the key grip and camera operator. The art director decides on sets, and is in charge of the construction crew. Finally, the key grip and the assistant director are responsible for safety on the set. Another key function is that of the location manager, who is responsible for permitting and for handing back the site in the right condition, something that some location managers do better than others.

At the studios, the environmental managers play a major role. Individuals at some of the major studios have put in place a wide range of environmental initiatives. The studios freely share information about their environmental practices with each other, and the environmental managers meet regularly. In most studios, the environmental department is very small, sometimes consisting of a single person who is also responsible for a wider range of government affairs. At least one of the studio environmental managers meets with each production crew during preproduction to discuss recycling on the stage, donating sets after use, how to plan for deconstruction rather than trashing of sets, etc. From that point on, it depends on the crew what they actually do.

An additional challenge is that the studios compete to rent out their sound stages to production companies (including their competitors). As a result, studios have limited ability to encourage environmental behavior by the production crews on the sound stages, as a studio that is perceived as being too difficult to work with will not be able to rent out their sound stages.

10.2.2 The Budget

A theme that recurred throughout the interviews was expressed by one respondent as follows: "The variables of production in the motion picture industry are corporate and creative, but the corporate outweighs the creative. It's all about making budgets not movies."

Within the studio system, there are three typical scenarios for funding movies. Some are 100 percent funded by the studio; others are coproductions, sometimes "single pot" where production costs are split 50–50 with another distributor; some are "split" where one distributor has the rights to the domestic market, another has the rights to foreign markets; and some are funded through sale of rights, where a single producer/distributor sells the distribution rights for individual territories.

At least one studio controls waste by never approving the first proposed budget: this is "our major means of control over waste versus efficiency in a production." Most budgets include a 10 percent contingency to cover unforeseen expenses, but production companies anticipate that, so there is always some manipulation of numbers going on. The 10 percent contingency is always spent, there is never a bonus (in practice) for completing a project under budget, and setting a higher contingency is also not desirable, as "then you look like you don't know what you're doing and you're a bad producer. The best thing is to spend right up to the budget."

A typical breakdown of a $50 million film budget would include $5–10 M for an A-list director, $15 M for an A-list actor and appropriate cast, and $2.5 M for an A-list producer. The remaining $22.5–27.5 M becomes the physical production budget, or the below-the-line part. If the budget needs to be cut, the producer will first start cutting shoot days, which cost about $150,000 per day for actors, crew, catering, vehicles, etc. This means removing pages from the script. Another option is to remove stunt sequences. Even if better environmental practices (such as energy conservation) lead to cost savings, those savings might have to be as high as $100,000–$200,000 for a larger production to actually change its behavior.

Transportation is often the second largest line item in a television budget (after talent), for example, $30,000 per day in Los Angeles for one particular show. The 30-mile zone is a key concept here: "centered at the old offices of the Association of Motion Pictures and Television Producers office at Beverly Boulevard and La Cienega Boulevard" in Los Angeles, it "defines a line past which union members of the Screen Actors Guild and the International Theater and Stage Employees must be paid per diem benefits" (Lukinbeal 1998, p. 71). The result is that this 30-mile radius "represents the most heavily filmed area in the world (Counter 1997), accounting for roughly 75 percent of all motion picture production and filming (Mosher 1997)" (Lukinbeal 1998, p. 71). The decision whether to shoot in Los Angeles or not is typically a trade-off between reducing costs (below the line) by moving elsewhere and the superior infrastructure available in Los Angeles. For that reason, made-for-TV movies are usually shot outside Los Angeles, while series, which depend more heavily on the availability of permanent infrastructure, are usually shot in Los Angeles. The production designer's part of the budget will typically be approximately 65–70 percent labor and 35–30 percent material.

10.2.3 The Schedule

Many productions are characterized by intense time pressure and chaos, as Coget (2004, pp. 56–57) describes:

> For the readers who have never set foot on a movie set, it is important to give a quick flavor of what it looks like. The first time one sets foot on a movie set can be daunting. Trucks full of equipment are scattered around in no apparent order. Crew members are hanging out every-where, fiddling with strange equipment, talking on walkie-talkies, marching in different directions, talking with other crew members in cryptic terms, or just waiting. It's impossible not to get a sense of chaos

looking at how these people work. One really wonders how they can coordinate their activity amidst such complexity. Yet, everybody seems to know what they are doing and seems at least minimally mindful of what others are doing. At given moments, such as when the AD [assistant director] shouts, "Quiet please! Rolling," followed by the director's quintessential "Action!," all conversation stops as each person quietly focuses on their part of the work and watches the action.

The director is the person orchestrating this "chaos," with the help of his/her DP [director of photography], AD and the key department heads, (The head of Art Department, the Costume Designer, the Set Designer, the Key Grip, in some cases the Special Effects Coordinator, the Stunt Coordinator, etc.). Following a director on set for five minutes quickly reveals an essential dimension of his/her job. At any given moment, the director is coping with a number of issues that need to be addressed. While walking around addressing issues—by observing and talking to his/her crew and cast—the director is bombarded by crucial appearing questions thrown at him/her by crew and cast members. Everybody seems to be fighting for the director's attention. Therefore, it is absolutely essential for directors to filter the stimuli that they gather on their own or that they are fed by their crew members so as to discern, select and prioritize what issues are most important for the shoot.

A consequence of this chaos is that the industry squanders human energy, although the large budgets of some major productions hide the waste of time and energy and inefficiency. Consistent with this, one interviewee, a costume designer, believes that the biggest waste is caused by fear: directors and producers are paralyzed and do not make choices until the last minute. The costume designer often does not receive a cast list or script until the night before, which means they cannot plan or budget properly, which leads to more waste and higher expense. As another interviewee, a line producer, noted, "Things take forever, then, all of a sudden ... we need it tomorrow." That this is not inevitable is shown by his experience with a particular film where the director did 4–6 weeks of rehearsals with the actors, well before the actual filming, and then shot for 10–12 weeks. This meant that each department had enough time to do research based on decisions that were made during rehearsal. As a result, the film came in under budget, despite the apparent expense involved in rehearsing. Singleton (1996, p. 86) confirms that rehearsing can save a lot of time during shooting and hence can also save money.

With this background on the motion picture industry, we now turn to various closed-loop supply chain examples we encountered.

10.3 Closed-Loop Supply Chains in the Motion Picture Industry

We start with some observations about recycling at the studios and on location in general, then discuss closed-loop supply chains for paper, film, and re-refined oil, and set materials.

10.3.1 Recycling at the Studios

At the studios, a considerable degree of recycling is in place. James (2000) describes this in detail, in a report produced for the Solid Waste Task Force, itself formed by the studios in "an effort to ensure that the industry contributes to the City's compliance with the Integrated Waste Management Act (IWMA) requirements of 50 percent diversion." Table 1 in James (2000) shows that, in 1999, 28,090 tons were diverted out of 46,007 tons generated, a diversion rate of 61 percent. Table 2 in James (2000) shows that the four Los Angeles–based studios even achieved a 69.7 percent diversion rate in 1999. Twentieth Century Fox averaged over 80 percent during the four preceding years (James 2000, p. 5). Warner Bros. has received several awards for recycling (Baker 1996).

The Alliance of Motion Picture and Television Producers (AMPTP)/Motion Picture Association of America (MPAA) Solid Waste Task Force has collected and published the solid waste diversion rates for the major studios since 1990. The Solid Waste Task Force includes Fox Studios, The Walt Disney Company, Paramount Pictures, Sony Pictures Entertainment, Universal Studios, Warner Bros. Entertainment Inc., Metro-Goldwyn-Mayer Studios Inc., and the West Coast broadcast and production centers of ABC and CBS.*

10.3.2 Paper

Paper consumption is a highly visible aspect of the motion picture industry. For instance, in 1993, Sony Pictures consumed 103 million sheets of paper, with scripts

---

* From "ENTERTAINMENT INDUSTRY SAYS "CUT!" TO PRODUCTION WASTE: Solid Waste Task Force Reports Total Recycling Rate of Over 68 percent," *MPAA Press Release*, April 21, 2005. This press release also explains that "The Solid Waste Task Force" (SWTF), comprising the major studios and television networks, was formed in the early 1990s, following the passage of the IWMA, to address resource conservation and reduce solid waste being sent to landfills. The SWTF Member Companies voluntarily implement waste diversion programs to reduce the environmental impact of solid waste, as well as assist the local government in meeting the mandates of the IWMA. Today, SWTF members meet regularly to collaborate on creating additional progressive environmental programs."

forming the largest portion of paper waste. Many copies of a revised script are distributed every day, often discarded unread, but by rethinking the distribution system, Sony reduced duplicates and unnecessary copies.* Several respondents indicated that recycling was common.

10.3.3 Film

Film stock is frequently recycled. Film Processing Corp. (FPC), now a subsidiary of Eastman Kodak, recycles film stock either into new plastic film base or for use as fuel.† FPC annually recycles more than 10 million pounds of film stock, of the 35 million pounds created annually.‡ Kodak does not charge for this recycling service, considering it part of the company's zero-landfill policy.§ FPC, under the leadership of Barry M. Stultz and Milton Jan Friedman, was awarded an Award of Commendation by the Academy of Motion Picture Arts and Sciences on March 4, 2000, in recognition of their pioneering role in film recycling.¶ One major factor underlying the studios' cooperation in film recycling and hence of FPC's business success lies in the antipiracy value of properly recycling old film stock, not just the environmental benefits.

Some efforts are in place to reuse film, or to reduce consumption altogether. Sony Pictures launched a program to reuse trailers, the film previews shown before the main feature in a movie theater. Theaters can send trailers to National Screen Service, which in turn will distribute these trailers to discount and second-run theaters to be reused. Only trailers that are too worn out are sent to FPC for recycling.†

At least one major studio requires the use of digital technology for 90–95 percent of all film projects, as they are easier to distribute than the traditional "dailies" and easier to archive, hence drastically reducing the amount of film stock consumed in the first place.

---

* "It's a Wrap: Hollywood Studio Spotlights Waste Reduction," *EPA Reusable News*, Summer/Fall 1995, pp. 6–7. http://www.epa.gov/epaoswer/non-hw/recycle/reuse/rnf5pdf.pdf, last accessed April 27, 2005.
† "Film Recycling Gets Reel," *EPA Reusable News*, Summer/Fall 1995, p. 6. http://www.epa.gov/epaoswer/non-hw/recycle/reuse/rnf5pdf.pdf, last accessed April 27, 2005.
‡ Purchasing.com, January 15, 1998; http://www.manufacturing.net/pur/article/CA109704; last accessed June 21, 2004.
§ Harry Heuer, director of health, safety and the environment for professional motion imaging; quoted in Purchasing.com, January 15, 1998; http://www.manufacturing.net/pur/article/CA109704; last accessed June 21, 2004.
¶ http://www.oscars.org/press/pressreleases/2000/00.01.10.html; last accessed June 21, 2004.

10.3.4 Used Oil and Re-Refined Oil, and Waste Tire Management*

Warner Bros. Entertainment Inc. operates a motion picture and television production and postproduction facility located on 142 acres in Burbank, California.[†] Production facilities include 34 soundstages and a 20-acre backlot. Warner Bros. has all of its own municipal services including a fire department, police department, parks department, sanitation, transportation, plumbing, and electrical departments. They employ thousands of people in a broad range of fields, including office work, fabrication, construction, and production, and are literally a "city within a city."

Warner Bros. has implemented numerous environmental programs such as their environmental purchasing program, a sustainable design and construction policy, an energy conservation program, and a green building philosophy. In the last 12 years, Warner Bros. has increased their waste diversion rate from 7 to 53 percent, recycling and donating 2983 tons of materials in 2003 alone. Annually, Warner Bros. saves $150,000 in disposal costs, generates $25,000 in revenue from recyclables sales, and has reduced energy consumption by over 6 million kW-h or $760,000.

In 2003, Warner Bros. diverted more than 2,900 tons of materials from the landfill, which resulted in waste hauling and disposal savings of $150,000 plus additional revenues generated from the sale of recyclables that totaled approximately $25,000. The environmental benefits of diverting these materials from the landfill and using them for beneficial uses has resulted in more than 700 MTCE (metric tons of carbon equivalent) of greenhouse gas emission reductions, saved more than 12,000 trees, and nearly 11,000 MBTU (million British thermal units) in energy savings.[‡] Additionally, Warner Bros. has reduced their energy use, which resulted in a savings of $760,000 annually.

Warner Bros. has switched their entire fleet of vehicles from virgin lubricating oil to re-refined oil. Since 1997, they have used re-refined oil in their fleet of over 400 vehicles including passenger cars, forklifts, and trucks. Although initially encountering some "hurdles," Warner Bros. successfully made the transition to re-refined oil without a single problem, and without additional costs or warranty issues. Warner Bros. has also implemented a used oil recycling program and returns all oil removed from their vehicles to be recycled and made back into re-refined oil. Warner Bros. conducts true closed-loop recycling with their environmentally conscious motor oil program.

---

* Original version contributed by Brenda Smyth, CIWMB, based on public information on file at CIWMB, including the California Integrated Waste Management Board WRAP applications from Warner Bros., 1993–2004.
† The background information about Warner Bros. was collected from http://www.wbjobs.com/?fromnav=movies, last accessed July 3, 2005.
‡ NERC (Northeast Recycling Council, Inc.) Environmental Calculator, December 2004.

Warner Bros. is one of the few businesses in the motion picture industry that have implemented a re-refined oil program, but they are among other California businesses and organizations that have implemented re-refined oil programs including Southern California Gas, the County of Los Angeles, the California Highway Patrol, Coca-Cola, United Parcel Service, the City of Sacramento, Waste Management, the U.S. Postal Service, Frito-Lay, CalTrans, the City of San Francisco, and Ventura County. Re-refined oil is even used on the NASCAR race circuit.

The environmental benefits of using re-refined oil are threefold: (1) used oil that was disposed of as a waste before can now be considered a renewable resource, eliminating the negative impacts of disposing oil to the environment; (2) re-refining oil extends the life of a nonrenewable resource, fossil fuels, by converting used oil into a marketable material that can be used, recovered, converted, and used again; and (3) 30 percent energy savings are realized because it takes roughly one-third of the energy to reprocess used oil than it does to refine crude oil into lubricant quality. Furthermore, the impacts of improper disposal of used oil can be devastating to the environment. For example, one gallon of used oil can contaminate 1 million gallons of drinking water.

10.3.5 Recycling on Location

In contrast to the relative success of recycling at the major studios, there are few or no recycling services catering to independent productions or to productions filming on location. Recycling costs money as it requires two dumpsters rather than one. For beverage containers, Hollywood Recycles is a free service for film and television productions filming within the Los Angeles area but away from major studio lots.* This program is funded by grants, which need to be renewed each season. Organizations such as L.A. SHARES, LooneyBins, and The ReUse People (TRP) all work with materials from locations, not just on-site at studios. Once a production unit leaves the confines of the studio complex, the logistics of reuse and recycling become more complex, and the volumes at any given location become much smaller, posing significant economic and organizational challenges for recycling on location.

The examples so far demonstrate that reuse and recycling applies to many product types in the motion picture industry. However, the materials used in making sets are likely to be what most people think of first when they consider recycling in the motion picture industry. We first describe set recycling in the industry in general, then focus on two case studies in more detail.

---

* For program goals and a list of current and past participants, see http://www.eidc.com/epg/ Hollywood_Recyles/hollywood_recyles.html; last accessed on April 25, 2005.

10.3.6 Set Recycling

Set recycling remains a challenge. It is usually cheaper for the art department to throw sets away rather than dismantle and reuse them, and indeed, most projects do not recycle sets because it is easier and more cost-effective to simply throw them away. When filming in the studio, there is a slight incentive to be sustainable, as it is cheaper to haul away clean wood than mixed trash; however, the cost savings is very small compared to the cost of a movie, and it depends on the specific crew (construction coordinator) whether they make use of this potential cost saving. (Recall the comment reported earlier from a studio executive that it would take $100–$200,000 in cost savings for a larger production to change its behavior.)

Each shot is its own project, and although the overall construction is carefully planned, the rough carpentry needed for each specific set is not, and once a shot is complete, the lumber is usually not kept for the next shot. Only some materials, such as apple boxes that are used to raise actors or act as steps, are reused. Some sets are stored for integrity. During filming, the "walls" pass from one department to another: they belong to the construction department until the camera starts running, then they belong to the grips for the duration of the shoot, after which they become part of the film, meaning that the integrity of the item is important in case the scene needs to be reshot later. They are carefully stored for this possible later reuse.

Some studios have been successful in internally reusing set materials used on-site. For instance, Warner's television programs reuse each other's sets while structural materials from films are reused in the company's own buildings (Baker 1996). The Disney Web site also refers to the storage of used sets and other materials available for rental.*

For materials that cannot be reused internally, some other programs exist to reuse set materials elsewhere. The CIWMB manages the California Materials Exchange (CalMAX), a free service for organizations to buy and sell used materials that would otherwise have been discarded. Recycling firms such as LooneyBins (see below) use CalMAX to help find users for leftover set materials.† TRP (see below) disassembles entire sets to salvage building materials for resale. L.A. SHARES takes donations of reusable materials from the business community, including most if not all studios, and redistributes these items free of charge to nonprofit organizations and schools.‡

Warner Bros. and Twentieth Century Fox collaborated to construct a database of over 750 nonprofit organizations (including L.A. SHARES) that could benefit from discarded materials in a program called Second Time Around (Baker 1996). For instance, parts of the sets from *Ocean's 11* were donated to the new

---

* See http://studioservices.go.com/production/backlot_services.html and http://studioservices. go.com/production/history_and_news.html, last accessed July 4, 2005.
† http://www.ciwmb.ca.gov/CalMAX/Connection/1998/Fall.htm, last accessed May 22, 2005.
‡ http://www.lashares.org/, last accessed April 27, 2005.

Natural Resources Defense Council headquarters building and other organizations (Tereshchuk 2003).* Similarly, the staircase in the Southern California Gas Company's Energy Resource Center was salvaged from the set of the Warner Bros. movie *Disclosure*.[†] Many items are sold through eBay.com, which can also be a source of materials; for instance, the Ferris Wheel used in *Lords of Dogtown* was bought through eBay. According to Ted Reiff, President of TRP, markets do exist for many materials, sometimes locally, sometimes in Mexico, or further; the key challenge lies in minimizing costs involved in handling and distribution.

Sets that are no longer needed can be recycled by LooneyBins, a waste-hauling company that sorts and recycles construction and demolition debris, achieving over 70 percent landfill diversion rates.[‡] LooneyBins recently received a $2 million low-interest-rate loan from the CIWMB to help it expand its recycling operations in Los Angeles.[§] Although LooneyBins does not operate in other countries, set building contractors in some other countries do reuse materials as it is profitable there; for instance, when labor is cheap, it can be economically viable to remove nails from wood to reuse it. Some materials, such as metal, can be sold after use; the art director's budget sometimes receives those funds, which can be used to cover other expenses within the art department, rather than returning to the overall production budget.

10.3.7 Set Recycling Case Study 1: Kenter Canyon Charter School Library[¶]

One of the more visually dramatic examples of movie set reuse comes from the Los Angeles Unified School District's new Kenter Canyon Elementary School Library. New Line Cinema released *Life as a House* in 2001. Those that saw the movie saw an attractive 1100-square foot Craftsman-style house being constructed, and unlike the majority of movie sets that are made from lighter weight and generally less durable materials, much of the structure was made from high-quality grade #1 Douglas fir. But, like the majority of movie sets, after the filming was completed, the set was to be demolished and disposed.

---

* Kristy Chew of CIWMB notes, based on conversations with NRDC staff, that the *Ocean's 11* sets were too big for them to use and were donated to other organizations; for example, the Watts Labor Community Action Committee received lighting equipment, ceiling panels, lobby furniture, computer racks, and cable conduit. The NRDC in Santa Monica did receive sofas from Warner Bros., although they were not used in *Ocean's 11*. From the *Ocean's 11* set, the NRDC building received some fluorescent lighting.
† CIWMB Publication #422-96-043, October 1996, available at www.ciwmb.ca.gov/Publications/GreenBuilding/42296043.doc, last accessed May 22, 2005.
‡ See under "recycling" at http://www.looneybins.com/, last accessed April 27, 2005.
§ http://www.ciwmb.ca.gov/Pressroom/2004/March/11.htm, last accessed April 27, 2005.
¶ Original version contributed by Kristy Chew, CIWMB.

The fate of the movie set was altered though when the movie's costume designer and also a parent of a child at the Kenter Canyon Elementary School envisioned the reuse of the structure. She contacted a fellow Kenter Canyon parent who was also an architect, who was impressed by the structure. The architect enlisted the help of another Kenter Canyon parent, an executive with a construction project management company.

The Kenter parents were able to work with New Line Cinema to donate the structure to the Kenter Canyon School. However, for a number of reasons the actual structure could not simply be picked up and moved to the school (e.g., the movie structure was smaller than the needs of the school, and the lumber needed to be re-graded and certified before it could be used in a public school). In the end, approximately 2000 cubic feet (ft^3) of lumber (e.g., sheathing, beams, and posts) from the 1100 square foot movie house were reused in the Kenter Canyon School library and 500 ft^3 of lumber were reused in its construction (e.g., concrete forms, window casings). The donated lumber is valued at about $35,000. The school and parents raised the additional funds needed to build the library and acquire the necessities to create the state-of-the-art facility that they desired, such as additional books, computers, and other library materials, and to fund a librarian position.

The school and parent volunteers overcame a number of physical and economic challenges associated with reusing the movie house for the Kenter Canyon School library. Prior to removing the structure from the movie shoot location, the school/volunteers had to provide proof of insurance and bonding. The Kenter parent with the construction project management company provided the insurance and bonding. Due to the movie studio's lease limitations, after the insurance was secured, the parents and volunteers had only three days to dismantle and move the movie house from the ocean-side shooting location.

Another major hurdle in the project was obtaining the permits and approvals from the State's Department of General Services, Division of the State Architect, the entity that reviews plans for public school construction projects to ensure that plans, specifications, and construction comply with California's building codes. For the lumber to be used as structural supports at the school, the wood stain that was applied for the movie had to be removed and then the wood had to be re-graded by a certified grader. The lumber salvaged from the movie set provided about half of the lumber used in the new 1700 square foot library, which was a significant cost saving.

Approximately 2500 ft^3 or 62.5 tons of lumber were salvaged from the *Life as a House* structure and diverted from the landfill. The reuse of 62.5 tons of dimensional lumber is the equivalent of reducing greenhouse gas emissions by 34 tons (MTCE)* by eliminating the need for virgin resources. Reusing 62.5 tons of lumber also saved energy, about 213 million BTUs, enough energy to run two average households for one year or 41 barrels of oil.*

---

* Northeast Recycling Council Environmental Calculator, December 2004.

The existing Kenter Canyon School's library, originally built in 1955, contained about 6000 books, and was less than 750 square feet in size; too small to adequately hold even one classroom of students. The new library, built in the same Craftsman style as the movie house, is 1,700 square feet, holds 11,000 books, and is able to accommodate up to three classrooms at a time. The library is now modern and comfortable—a real gem to the school's 450 students.

The energy, enthusiasm, and fund-raising that went into the new library have allowed the school to have a librarian and provide new and expanded library programs and technology for the students, whereas in many schools, library programs have decreased due to budget cuts. The enthusiasm for learning by the students and the new and expanded programs offered by the library has increased hand in hand. The new library is now hosting "story times," guest authors, book groups, homework clubs, study groups, and special events for the students.

10.3.8 Set Recycling Case Study 2: The ReUse People Salvage the Sets from "The Matrix" 2 and 3*

TRP is a nonprofit organization that started in San Diego in 1993 and began operating in Alameda County in 1999.[†] The two sequels to *The Matrix*, known as *The Matrix Reloaded* or *The Matrix 2* and *The Matrix Revolutions* or *The Matrix 3* were both released in 2003, by Warner Bros. After the success of the initial release, the budget for *The Matrix 2* was an estimated $127 M, and a worldwide gross of $736 M; *The Matrix 3* had a $110 M budget and grossed $424 M worldwide.[‡] Parts of both films were shot at three sets and on the streets in Oakland and Alameda Point.[§] The sets were large. For instance, the cave set consisted of 90 tons of material: wood and polystyrene blocks. The tenement set consisted of 300 tons of material, representing eight building fronts. The freeway set consisted of more than 7700 tons of concrete, 1500 tons of structural steel, and 1500 tons of lumber. As a result of a joint project between Warner Bros., the city of Alameda, the Alameda County Waste Management Authority, and TRP, 97.5 percent of all the set material was recycled.

---

* This section was written with assistance from Ted Reiff, President, The ReUse People, who is the source for much of the information included here.
† See http://www.thereusepeople.org/ for more information. See also Benefits of Regional Recycling Markets: An Alameda County Study, California Integrated Waste Management Board, 2003, pp. 19–20, accessed at http://www.epa.gov/jtr/docs/ca_98report.pdf on January 5, 2005, for more information on The ReUse People's salvage and retail operations.
‡ According to www.imdbpro.com, accessed on January 5, 2005.
§ Dennis Rockstroh, San Jose Mercury News, December 2, 2001; reproduced at http://www.thereusepeople.org/inside.cfm?p=VelvetCrowbar&recordID=92, accessed on January 5, 2005. Much of the information in this section was obtained from this article.

TRP dismantled the set piece by piece, and handled processing and distribution of the salvaged materials. According to Rockstroh*

- "The lumber was sold to a company that builds housing for low-income families in Mexico. Thirty-seven truckloads went south, and three truckloads went to The ReUse People's yard for resale to the public.
- One hundred percent of the steel was used as is.
- Some 48 fire escapes were sold to area contractors along with more than 60 decorative moldings.
- The polystyrene blocks were sent out for use in insulation material.
- And 3.9 miles of k-rail from the mystery freeway was broken up, crushed, stored on-site, and eventually sold off as class 2 base rock."

As of December 15, 2004, the last of the fire escapes at TRP's warehouse was sold.

What makes this achievement all the more remarkable is that the cost to the production company is unlikely to have been significantly higher than it would have been using a more traditional demolition firm, as it is unlikely that TRP would have been selected for the work otherwise. In the Fall of 2001, the California Film Commission awarded Warner Bros. the "Humanitarian Award" for its environmental stewardship.†

TRP's workforce of 18 people worked 124 days to complete the project. According to the Alameda Waste Management Authority, the 11,000 tons diverted from the landfill represented 10 percent of the total annual solid waste stream for the city of Alameda.

Had a traditional demolition company completed the project instead of TRP, the expanded polystyrene, plywood, oriented strand board, and truss joists would all have been landfilled. These materials, due to their composition could not be ground up for ground cover or cogeneration as we do for small pieces of clean lumber.

Ted Reiff adds that the contract with TRP was signed after the filming was finished, and TRP did not have the chance to consult with the production people before construction began to suggest ways to improve the time and cost of salvaging. For instance, if standard k-rail had been used or made, there would have been no need to breakup and crush the k-rail that was eventually used. This would have saved over $30,000 and approximately two to three weeks. Throughout this process, TRP's contact was with the production's location manager.

For reasons outside their control, the dismantling of the freeway sets was started by another organization than TRP; TRP reduced its contract price in lieu of this dismantling work. However, the work was performed so poorly that TRP spent even more time correcting the work once they got on the job. In future contracts,

---

* Rockstroh, cited above.
† *In Business*, 24:2, March/April 2002, p. 4.

TRP will not allow others to dismantle sets without their supervision and guidance. Other than this slight glitch, TRP's total crew days on the project were only 3 days over its forecast of 121 days.

10.4 Conclusions

We have seen that even in a primarily information-based industry such as the motion picture and television industry closed-loop supply chains do exist. The degree to which closed-loop supply chain practices are implemented varies widely from project to project.

Some of our interviewees consider the motion picture industry to be unique, while others resolutely contradict that. This carries over to the environmental practices within the industry: those who believe the industry is unique are more likely to feel that the planning, construction, disposal, and energy-efficiency approaches that work in more traditional industries cannot, almost by definition, work in the motion picture industry. Others, though, view the motion picture industry as much more similar to traditional manufacturing industries than many would like to admit, implying that there is little excuse for not adopting the same principles as firms in other industries do. In some respects, several individuals within the industry have indeed implemented programs that firms in any other industry would be proud of. In others, however, the industry is decidedly conservative, both in terms of adopting new programs and in terms of discussing them with the public.

The interviews we conducted provide a rich and varied view of the many obstacles, some real, some imagined, facing widespread adoption of environmental practices (including CLSC practices). Clearly, educating individuals about the environmental choices they have is the key, as well as continuing to find and publicize environmental practices that are cost-neutral or even cost-beneficial.

Some environmental opportunities require behavioral change within the industry. Most of all, the prevailing tendency within the industry is to operate in a stop-go mode, both at a large scale and at a small scale. At the larger scale, very little happens for lengthy periods while a project is in its early stages, but when key agreements with financiers or talent fall into place, it switches into a full-speed-ahead mode when everything needs to happen as fast as possible. At the smaller scale, during production, an entire production crew needs to be present on a set even when most of the time nothing appears to be happening, just to ensure that everything can happen quickly during the critical moments when actual filming occurs. Several of the interviews indicate that more careful planning, of the overall project and of actual shooting, can at least in part reduce the uncertainty and resulting tension, which in turn helps by giving individuals the time to consider and implement more environmental choices. This applies to recycling set materials too: deconstruction usually takes longer than demolition, and deconstruction can be done more efficiently if the sets are constructed with deconstruction in mind. If one starts thinking about deconstruction only after the sets have been built and the shooting is finished, it is far less likely to be economically feasible than if it is designed into the process from the beginning. This is no different than in any other supply chain: products designed with end of life in mind are easier to disassemble and reuse, making closed-loop supply chains more likely to be profitable.

Acknowledgments

Many individuals provided valuable assistance during the original study. Dr. Richard P. Turco, Professor of Atmospheric Sciences, UCLA, Founding Director of the UCLA Institute of the Environment, was the other principal investigator for this project. The student team members, who performed many of the interviews, were Joanna Hankamer, Shannon Clements, and Jeannie Olander; other students who contributed in various ways were Fatma Cakir, Patricia Greenwood, Penny Naud, Kimberly Pargoff, and Michael Rabinovitch. Professor Barbara Boyle, Professor Mary Nichols, and Gigi Johnson generously shared their contacts and insights. Many individuals generously shared their time by agreeing to be interviewed by the team. Throughout the project, the UCLA team received exceptional assistance and support from the contract managers at CIWMB (Judith Friedman, Brenda Smyth, and Kristy Chew).

References

Baker, N.C. Warner Bros.: Loony about waste reduction, *Environmental Management Today*, 7(3), July–August 1996, 9.

Coget, J.-F.A.H. Leadership in motion: An investigation into the psychological processes that drive behavior when leaders respond to "real-time" operational challenges. PhD dissertation, UCLA Anderson School of Management, University of California, Los Angeles, CA, 2004.

Counter, D. President, Alliance of Motion Pictures and Television Producers. Personal correspondence to Chris Lukinbeal, December 10, 1997 (Cited in Lukinbeal 1998).

DeFillippi, R.J. and M.B. Arthur. Paradox in project-based enterprise: The case of film making. *California Management Review*, 40(2), Winter 1998, 125–139.

James, W.M. Waste Diversion Assessment: Major Motion Picture and Television Studios. Report prepared for the City of Los Angeles, Solid Resources Citywide Recycling Division, through a grant to the Entertainment Industry Development Corp (LA Film Office), Los Angeles, CA, 2000.

Jones, C. *The Guerrilla Film Makers Movie Blueprint*. Continuum, London, U.K., 2003.

Lukinbeal, C. Reel-to-real urban geographies: The top five cinematic cities in North America. *The California Geographer*, 38, 1998, 64–78, accessed through http://geography.sdsu.edu/Research/Projects/Film/Lukinbeal98.pdf on April 20, 2005.

Mosher, L. Librarian, California Film Commission. Personal correspondence to Chris Lukinbeal, November 12, 1997 (cited in Lukinbeal 1998).

Scott, A.J. A new map of Hollywood: The production and distribution of American motion pictures. *Regional Studies*, 36(9), 2002, 957–975.

Singleton, R.S. *Film Budgeting (Or, How Much Will It Cost To Shoot Your Movie?)*. Lone Eagle Publishing, Hollywood, Los Angeles, CA, 1996.

Storper, M. The transition to flexible specialisation in the US film industry: External economies, the division of labour, and the crossing of industrial divides. *Cambridge Journal of Economics*, 13, 1989, 273–305.

Tereshchuk, D. At Warner Bros., environmentalism plays a leading role, *AOL Time Warner Keywords Magazine*, April 2003, p. 7. http://mainegov-images.informe.org/governor/baldacci/news/events/docs/KW200304_April_2003.pdf, last accessed April 27, 2005.

U.S. Department of Commerce. The Migration of U.S. Film and Television Production. 2001. Available at http://www.ita.doc.gov/media/migration11901.pdf; last accessed April 25, 2005.

Chapter 11

Reverse Supply Chain in Hospitals: Lessons from Three Case Studies in Montreal

Rajesh K. Tyagi, Stephan Vachon,
Sylvain Landry, and Martin Beaulieu

Contents

11.1 Introduction

In recent years, the general population and the media have increased pressure on organizations around the world to adopt practices in line with environmental stewardship. The health care sector* does not escape this public scrutiny (Gilmore Hall 2008; Messelbeck and Whalley 1999) as it continually faces environmental challenges particularly from the volume of waste generated from its operations. Over the years, hospitals have increased the utilization of single-use, disposable supplies and materials leading to a substantial growth in the amount of waste generated (Fisher 1996).

A key indicator for waste management in a hospital is the weight of waste per bed. While historically this indicator suggests a variance in the amount of waste generated on average per bed from 1.5 to 6.8 kg (Flinders Medical Center 2000; Gilden et al. 1992), more recent statistics indicate that the Canadian hospitals' average is about 5.5 kg (Environment Canada, 2009). The variability from one hospital to another can be explained by the type of facility (i.e., extent and variety of treatments, surgeries, or specialties) as well as by the proportion of disposable to reusable items (Curtis and Mak 1991). It is noteworthy that about 15–20 percent of the waste generated in a hospital are biomedical (Chandra et al. 2006); this amount comprises, however, 80 percent of recyclable or reusable materials (paper, cans, bottles, and packaging) that were contaminated (Weir 2002).

Due to the diversity of activities (e.g., emergency, surgery, or oncology), the management of waste is a major and relatively complex undertaking for a hospital. This challenge is compounded by governmental agencies that have developed and enacted policies and regulations regarding waste management in hospitals. For example, such governmental actions to address the waste problem in hospitals are getting particularly stringent in industrial countries such as the United States (EPA 2005), the United Kingdom (Tudor et al. 2005), and South Korea (Jang et al. 2006).

As a provider of medical service to end-consumers (i.e., patients), hospitals constitute the juncture point between the forward and the reverse flows of materials (disposal, recycling, reuse). As such, they are the last commercial buyer and the responsible party to initiate the reverse flow. Therefore, their role in closing the loop is fundamental.

---

* The health care sector comprises several providers including ambulatory services, hospitals, and nursing and residential care facilities. This chapter emphasizes particularly the hospitals.

The reverse supply chain in hospitals* has not just been viewed as a component of environmental stewardship but in some cases it is foremost adopted as an approach to contain or control operating costs and to generate revenues (Chandra et al. 2006; Geiselman 2004). As put by one hospital manager "the bottom line is you can either spend up to $50 per ton to get rid of your [waste], or you can make more than $50 per ton by recycling it" (Geiselman 2004). For example, the Albany Medical Center (Albany, NY) distills waste chemicals for reuse, saving $250,000 per year in chemical disposal and purchasing costs (Kaiser et al. 2001). Another example is the Beth Israel Medical Center (New York, NY) that has implemented a program to rigorously reduce the amount of solid waste going into the designated "red bags" for biohazardous waste (Kaiser et al. 2001). This effort saves the hospital $900,000 per year in disposal costs by reducing the amount of biohazardous waste that must be treated.

Over the past decade, some organizations have established programs and implemented practices addressing the issue of closing the loop in the health care sector (Birk 2008; Donston 2007; Hamilton 2008; Messelbeck and Sutherland 2000). Unfortunately, despite its growing popularity, the reverse supply chain associated with the health care sector is not extensively discussed in the literature. There is no systemic model or framework that has been proposed in the literature to guide managers with regard to reverse material flows and its management. Furthermore, aspects of the implementation of a closed-loop program are rarely presented and discussed. Using the elements of a supply chain structure proposed by Chopra and Meindl (2007), a conceptual framework is developed and substantiated through anecdotal evidences from the literature and interviews with three organizations that have addressed these issues in their operations.

This chapter contains four additional sections. Section 11.2 explores the main logistical issues specific to hospitals and their reverse supply chain. Section 11.3 introduces a conceptual model comprising the different elements of the reverse supply chain structure. Section 11.4 provides specific examples related to the conceptual models from the literature and the interviews conducted in different organizations. Section 11.5 presents some managerial implications, particularly associated with the implementation and the different lessons learned from the literature and the interviews.

11.2 Reverse Supply Chain in Health Care Sector

In traditional forward supply chains, health care institutions are not just another link as logistical activities are dispersed throughout a set of heterogeneous activities

---

* The reverse supply chain and the notion of closed-loop supply chain have been considered synonymous in the literature (Kocabasoglu et al. 2007). In fact, French and LaForge (2006) have used reverse logistics and closed-loop interchangeably.

(e.g., emergency room [ER] versus orthopedic clinic). In fact, two main logistical factors have hindered the possibility for the hospital to fully integrate its supply chain: (1) hospitals are the convergence point for a wide variety of products that support health care directly (e.g., medical supplies and pharmaceutical products) and indirectly (e.g., linens, food, stationery, maintenance products) and (2) the diversity of material flows bring with them a variety of stakeholders interacting within health care institutions, which in turn increases the complexity of the logistical management. Often these stakeholders, most notably the nursing staff, have neither the expertise nor the resources to effectively and efficiently manage logistics activities (Landry and Beaulieu 2007). To have a coherent discussion, the focus in this chapter is put on the important segment of waste and byproducts from the daily operations. Reverse supply chains in the health care sector have inherited the complexity generated in the forward supply chain.

11.2.1 Type of Waste Streams

The U.S. Environmental Protection Agency (EPA) has determined three major solid waste streams: (1) general, (2) biohazardous, and (3) hazardous (EPA 2005). The latter two are usually highly regulated in developed countries while the first category has been the target of intense public scrutiny in recent years (Messelbeck and Sutherland 2000).

General wastes (also known as municipal waste) are very similar to wastes that are generated by any other mass service providers like hotels, shopping centers, or large office buildings. Hence, the general waste includes material like discarded packaging, paper, plastics, and foam cups and containers. As such, several tools and guidelines are available to the managers of health care facilities to implement a closed-loop program (EPA 2005). However, there is a particular challenge to health care managers as they need to assure the "cleanliness" of general waste. The general waste stream is often commingled with the two other streams making its management more difficult and complex.

Biohazardous wastes are generally characterized by the presence of contamination. Such contamination can come from infectious agents or human blood. Human pathological elements such as organs, tissues, or body parts are also included in the biohazardous waste. Finally, potentially harming wastes are considered biohazardous (e.g., sharps). By its very nature, this stream of waste is tightly regulated making it difficult to establish a comprehensive closed-loop program; however, as seen later, there are some possibilities to avoid disposal after the use of the products.

Other hazardous wastes that are not bio-contaminated but pose some risk through their handling or release to the environment constitute the third stream. In particular, spent chemicals (e.g., some chemotherapy drugs and formaldehyde), highly corrosive material, and materials with toxic content above the regulatory threshold comprise this waste stream. Often the main challenge within this stream is the management of pharmaceutical products within the hospitals as some of these products have tight expiration periods.

11.2.2 Approaches for a Reverse Supply Chain Strategy

A health care organization striving for environmental stewardship while aiming for operational efficiency can address the three waste streams using the well-known reduce–reuse–recycle (3Rs) approach. Although there are several challenges to apply all or parts of the 3Rs for each of the waste streams, the application of the 3Rs remains a cornerstone of a reverse supply chain strategy as shown in Figure 11.1. Because hospitals are service providers, the production of medical supplies and related materials (as well as the innovation incorporated in these goods) are conducted upstream by the suppliers. This is an important feature of the health care sector, and the efforts in implementing the 3Rs often rely on the interaction with suppliers or outside organizations. This is particularly true for waste reduction efforts that emerge regularly from suppliers' innovation or a close collaboration between the suppliers and the health care industry. Surgical custom packs that came into the market in the early 1980s are a good example of a solution based on buyer–supplier collaboration (DeJohn 2004). Such a custom pack eliminates the need for instrument's individual packaging reducing significantly paper, foam, and plastics wastes. Another example would be mercury-based thermometers, which can be phased out from health care organizations because substitutes such as the digital thermometers were brought into the health care industry by medical supplies vendors.

Table 11.1 presents other examples of the application of the 3Rs to the different waste streams. It is noteworthy that despite the tight regulation some health care organizations manage to find ways to reuse biohazardous waste. However, it is also

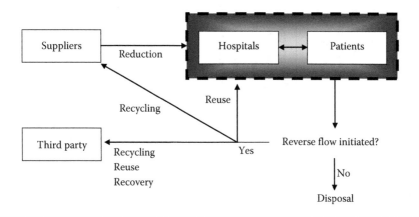

Figure 11.1   Approaches for reverse supply chain in health care.

Table 11.1   Examples of the 3Rs Applied to Health Care Operations

Waste Stream	Reduce	Reuse	Recycle
General	• Eliminate excessive packaging. • Use 55-gallon directly supplied drums for cleaning materials instead of multiple disposable small containers.	• Use dishes and cutlery that are durable rather than made of disposable plastics. • Use reusable ink cartridge for printers.	• Collect paper and paperboard (for third-party recycling). • Ensure the collection of non-contaminated plastics for recycling. • Install bins for beverage containers.
Biohazardous	• Work with suppliers to aid customized surgical package: avoiding unused devices going to waste. • Consider switching from disposable to reusable medical instruments	• Use washable/reusable linen for bedding. • Donate clean, unused operating room (OR) supplies for reuse overseas. • Use cloth diapers (reuse the worn diapers as cleaning rags).	• Regulatory context in most industrial countries limit considerably the possibility to recycle biohazardous waste.
Hazardous	• Substitute mercury-based thermometers for digital thermometers whenever it is possible. • Streamline the pharmacy inventory and operations management to avoid obsolescence. • Examine the possibilities for vendor-managed inventory in the pharmacy or for chemicals.	• Reuse batteries that are retrieved from critical medical devices by putting them into noncritical equipment. • Distill applicable solvent for recovery and reuse.	• Recover and recycle spent photographic fixer solution. • Collect and ship aerosol cans to a recycler. • Collect and ship lithium-based batteries.

Sources: EPA, *Profile of the Healthcare Industry*, U.S. Environmental Protection Agency, Washington DC, 2005; Rau, E.H. et al., *Environ. Health Perspect.*, 108, 953, 2000; authors' experience.

recognized that recycling biohazardous waste is basically nonexistent because of proscription by environmental and public health regulation (Rau et al. 2000).

New technologies allow automatic packaging and labeling of solid oral medications according to the needs of a hospital care units. The main attribute of such technologies is a significant reduction in products mistakenly ordered or over-ordered (Kratz and Thygesen 1992). Such a reduction in ordering mistakes translates into less medication-related wastes.* It is noteworthy that sound inventory management can generate significant reduction in packaging wastes. For instance, in some health care organizations corrugated paperboards are not sent to the care units, but are rather kept at the organization's receiving store, which delivers to nursing units separately instead of case loads. Moreover, a Montreal-based hospital implemented an exchange cart system between the vendor and the hospital's dialysis unit where the bottled solutions are delivered in reusable containers. Finally, using a two-bin replenishment system and its built-in stock rotation reduces product wastage generated by expired items (Landry and Philipp 2004).

Despite all the anecdotal evidence and the initiatives of several health care organizations, there still is a need for a comprehensive and systemic framework that can guide health care managers to craft and implement a closed-loop program.

11.3 Reverse Supply Chain Structure

To provide some guidance that can be used by managers wanting to develop a closed-loop program, we propose a conceptual framework related to structural elements in a reverse supply chain. The framework is inspired by the work of Chopra and Meindl (2007) that presents a conventional or forward supply chain structure. Originally, Chopra and Meindl's model consisted of four managerial drivers: facilities, inventory, transportation, and information. The reverse supply chain structure proposed here builds from their work but is adapted to a hospital setting. One of these adjustments is the nature of these drivers. Although the reverse supply chain structure presented in Figure 11.2 comprises the same number of drivers (i.e., four), they are slightly different; the four drivers of the reverse supply chain are facility, handling, the ease of access, and sourcing.

The inventory driver of conventional supply chain is replaced by the ease of access. In the fast-paced hospital environment, the system should facilitate the employees' involvement and participation (and for that matter other hospital's stakeholders such as the suppliers and patients) in the reverse supply chain. The perception of the employees regarding the ease of use of the system is fundamental in the effectiveness of any system that aims to initiate the reverse flow of material. Therefore, the ease of access includes the support provided by the hospital including clear return procedures, proper training, and various infrastructures.

---

* This is particularly true when rigidities exist within the health care organizations that prevent a mistaken ordered medication to be redirected to another department or patients.

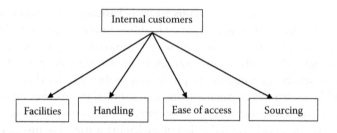

Figure 11.2   Reverse supply chain structure.

Another difference with the conventional supply chain consists in replacing transportation by "handling" as it appears that the logistical task of manipulating and moving wasted materials within the organization becomes key in the success of a closed-loop program.

It is also noteworthy that the responsiveness–efficiency spectrum, proposed in the original framework as a major contingency factor in deciding on the conventional supply chain structure (Chopra and Meindl 2004), is not as relevant for a reverse supply chain particularly in a hospital. That is explained by the fact that the management of return flows is not subjected or affected as much by the degree of uncertainty implied in the service delivery. For example, an ER at a hospital faces a very high uncertainty of customer demand. The demand is unpredictable in terms of timing, volume, and types of services required. On the other hand, the waste generated from the ER is more predictable with volume and mix that should be less volatile than demand for the service in the ER.

Finally, it is important to distinguish between the different types of customers that are involved in both the forward and the reverse supply chains. Generally, the customers remain the same between the forward and the reverse supply chains (e.g., manufacturer versus retailer versus end-consumers), but this is not necessarily the case here in a hospital setting. In a forward supply chain, the structural decisions are made with the patients in mind. In fact, the responsiveness–efficiency spectrum is of high importance in structuring such a supply chain. However, the reverse supply chain structure should be constructed with the employees in mind. Thus, the immediate customer for the forward and the reverse supply chains are not the same. In a hospital setting, the direct customers for the reverse supply chain are internal to the organization as they have to initiate the reverse flow of the material.

11.3.1 Facility

The facility driver refers to locations where returned or reused products are collected. Where the facility is located has direct impact on the overall cost of the system. Facility is usually decided at the design stage and has long-term strategic implications. The following questions should be answered while designing this driver: How much space is available, especially for the compactor? Is the facility located in-house or at a third-party location? Where is the facility located exactly within the building? During our interviews, we noted that even in hospitals that were built within the last five years, the issue of space is a very real one.

11.3.2 Handling

Handling refers to the segregation, containerization, and internal transportation of spent materials. The relevant questions while making decisions for this driver are the following: How waste streams are segregated? What is the extent of segregation? What are the primary means of internal transportation used? How segregation is practiced using color coding, classification schemes? How segregation practices are communicated hospitalwide? Hospitals face threefold challenges while designing this driver: the extent of segregated space, reorganizing tasks, and providing training to employees to enhance participation in segregation and providing safe and secure internal transportation within the hospital. Hospital environment poses a challenge for internal transportation and appropriate locations of containers. Containers placed in inappropriate locations can generate higher volumes of waste of the wrong kind. Some hospitals provide separate elevators for the transportation of wastes.

11.3.3 Ease of Access

The ease of access was considered an important driver based on qualitative research for reverse supply chain. The ease of access describes access policies and proximity to various facilities. This driver has not been explicitly reported and discussed in literature. We consider this driver to be equally important and often a deciding factor in the success or failure of a reverse supply chain system. This driver, which measures the self-reported perception of the customers, refers to the ease of access policies for the customer, access channels provided, and the availability of information about this channel.

A hospital needs to have clear policies and methods from customers' perspective that guides the reverse flow. Hospitals will send back more products in reverse flow and segregate more if the system makes it easier to do so. However, if the process of returns is difficult or nonexistent, there is a higher chance that the employees will lose their interest in the system. It is important to design waste collection frequency in such a way that containers are not overflowing and locations of containers should be clinically convenient. Some drug companies make it easier to return unused drugs by taking unused drugs back with no questions asked. In British Columbia, Canada, brand-owners of pharmaceutical products are responsible for the management of their product, including the collection of leftover products. The program was started in 1996.

11.3.4 Sourcing

The sourcing driver is important for any hospital considering the implementation of closed-loop supply chain. The sourcing driver includes decisions about supplier management and sourcing or outsourcing of the various components of the reverse supply chain structure. Some relevant questions while designing sourcing are the following: Should there be a few suppliers or many suppliers? How much supplier involvement is necessary in decision making? How are suppliers selected? How often is supplier selection reviewed? Sourcing decisions are important at hospitals and could differentiate hospitals from their competition. Hospitals receive and process supply from variety of sources including pharmaceutical companies, biomedical firms, food and beverage firms, and other equipment suppliers. Some examples of purchase practicing are selecting products with less packaging, digital imaging techniques such as the x-ray film technique, replacing traditional selecting drugs with longer expiration date, replacing polystyrene foam dishware with washable ceramic dishes, selecting products without preservatives, managing supply of drug samples, matching dosage pattern with the usage pattern, and using a reverse distributor to return expired drugs. Hospitals are increasingly putting an emphasis on environmental responsiveness of a supplier during the selection process. Hospitals are asking suppliers about take-back packaging and recyclable packaging. An example of such a practice is purchasing syringes packaged in paper/plastic wrap rather than in a rigid plastic tube. In fact, a recent American survey indicates that 80 percent of hospital executives would move away from their current suppliers if another supplier can offer a larger selection of sustainable products (Hale 2008).

11.4 Evidences from Three Montreal-Based Hospitals

To substantiate the model presented in the previous section, three hospitals in the Montreal, Quebec (Canada), area were selected. These hospitals offer a good range of size and complexity and will allow identifying some key implications for managers.

11.4.1 Observations from the Case Hospitals*

The first hospital (Hospital A) comprises three main buildings with 1150 beds. It has a long history in Montreal as their oldest building is more than 100 years old. Interestingly, Hospital A is in the process of building a brand new facility. The first conflicting point in building a new facility is space. Architects and project sponsors want to optimize the space with regard to care units and patients' comfort that puts space for recycling bins, segregation stations and rooms, and storage space at

---

* To maintain the confidentiality of the respondents, the name of each hospital remains anonymous.

premium. The facility's related issues are also apparent in the existing buildings where the administration and the staff would like to do more in managing spent materials but the required space is just lacking. For instance, the recycling efforts would greatly gain from a compactor, but there is simply no space in the existing facility to set up such piece of equipment.

Another challenge faced by the managers in Hospital A is the fact that first-line employees, despite being enthusiastic about the principles of the reverse flow, show signs of resistance. For instance, most of the areas in the hospital (e.g., admission/registration desks, nursing posts) are equipped with dual trash cans that can collect recyclable waste separately from the rest of the garbage, but the employees are not taking the time to empty the recyclables into a central depository. Without this first manipulation and material handling task, the effectiveness of the reverse supply chain is compromised. This resistance that prevents initiation of the reverse flow of waste can also be explained by low level of ease to access. Despite their long history in the area, no real collaboration is taking place with the suppliers.

Hospital B is a medium-sized hospital with 270 beds and is mainly housed in a relatively new building that is only ten years old. Even with a new building, recycling and other waste management initiatives were difficult because of a lack of space. Hospital B's managers would also have liked to install a compactor and an adjacent storage space, but they are not able to find the proper place on the premises. Hospital B manages 19 specialized facilities, covering clinical and long-term health care facilities. Some facilities are served by a private service provider, and others are served by the local municipal services. The choice between a private service provider and municipal services is driven by the volume handled and local municipal regulations. This hospital also considers that initiating reverse flows directly in the patient rooms are dangerous from a hygiene perspective, which limits their efforts within the common areas and different administrative units or stations throughout the hospital. Therefore, a comprehensive segregation and related reverse flows would necessitate more handling during the front-end of the reverse supply chain. From a sourcing perspective, very little efforts are made besides environmental criteria on packaging and specific products. The third hospital (Hospital C) is also a medium-sized hospital with 250 beds and a 50-year-old building. Hospital C's managers have also expressed the need for a compactor. From a handling perspective, a comprehensive recovery system put in place for the entire organization necessitated the use of a third-party contractor for the picking of waste throughout the hospital. However, personnel issues (i.e., unions) obstructed the implementation of such a system creating inefficiency in the removal operations. Hospitals A and B have already introduced sustainable criteria to reduce waste at source.

11.4.2 Discussion

The three observed hospitals reveal some common challenges and practices that can be synthesized. Issues related to the facility element of the reverse supply chain structure can be summarized in one word: space. The management of space links two managerial drivers, facilities and handling, presented in the framework. To initiate properly the reverse supply chain and to avoid contamination between waste streams, more space is needed for multi-bins stations, separated storage space, and the need of specialized equipment (e.g., a compactor). Even with a clean slate (i.e., the blueprint of a new building with Hospital A), space for reverse supply chain activities competes with others, and closer to the "core business activities." Actually, managers from two hospitals have mentioned explicitly that the core business of a hospital is to provide care not to recuperate spent materials or to increase the diversion rate to the landfill.

The interviews with the managers also suggest that the resistance to initiate the reverse flow was a critical factor particularly in two of the three hospitals. In one hospital, the issue was about employees considering the extra "handling" of material to be burdensome and in the other hospital the unionized workforce opposes the use of a third party to collect and manipulate the wasted materials. That is quite important as it suggests that even with greater ease of access to the reverse flow system, personnel issues can remain a major roadblock. This roadblock illustrates the importance of handling as a managerial driver. It also shows the interrelation between the four drivers put forward in the framework developed in the previous section.

It is noteworthy that the three hospitals do not leverage more the supply network to find new ways to reduce waste at the source. Besides marginal purchasing policies the sourcing driver does not really contribute to the reverse supply chain efforts in the interviewed hospitals.

11.5 Managerial Implications and Challenges

Managerial implications are discussed in two broad categories representing insights from the framework and learning from the field. Ideally, a rigorous validation of the framework requires extensive field data and statistical analysis. However, in this chapter, we provide a preliminary validation of the model based on secondary literature analysis and field interviews. As discussed in Sections 11.1 and 11.2, the literature describes the importance of facility, sourcing, and segregation decisions, individually. The reverse supply chain structure (presented in this chapter) provides an integrated framework to aid in decision making. The decisions embedded in the framework also represent the critical steps in waste stream management process at a health care facility. For example, the World Health Organization WHO's Web site (http://www. healthcarewaste.org/en/127_hcw_steps.html) on health care waste management describes the management of waste streams into eight distinct steps. These eight steps are waste minimization, waste generation, segregation and containerization, intermediate storage, internal transport, centralized storage, external transport, and treatment and final disposal. The four elements of reverse supply chain structure capture the essence of this disposal process. The sourcing driver relates to the steps of waste minimization and waste generation that can be used for source reduction at the planning stage. Handling driver is directly related to segregation, containerization, and internal transport. The centralized facility, representing the facility driver, is another important driver of this structure. The ease of access driver represents the acceptance and proximity to the processes of return, reuse, and recovery.

Our field interviews also support the importance of the elements of the framework. Interviews show that a hospital at the design stage has a long wish list with competing priorities. A closed-loop supply chain approach requires involvement from both management and other stakeholders; therefore it involves long delays before reaching a consensus. As discussed in Section 11.4, the hospitals emphasized the need for suitable facility and handling methodologies while designing a new closed-loop system. Education and communication is particularly important when it comes to properly managing the disposal of infectious medical waste. Some legacy systems and methods are difficult to change at an existing hospital and require top management intervention. As waste management is not a core business of a hospital, a waste management program is driven by the imposed regulations and stakeholders' motivation. The hospitals develop a recycling program especially to prepare health care organizations for eventual governmental policies and to project a good public image.

This study has several limitations. Insights gained in this study are based solely on interviews conducted at three hospitals and secondary literature. The framework presented in this chapter has not been validated.

References

Birk, S. (2008), An issue that can't be contained, *Materials Management in Health Care*, 17 (5), 42–44.

Chandra, H., K. Jamaluddin, L. Masih, and K. Agnihotri (2006), Cost-benefit analysis/ containment in biomedical waste management: Model for implementation, *Journal of Financial Management & Analysis*, 19 (2), 110–113.

Chopra, S. and P. Meindl (2007), *Supply Chain Management: Strategy, Planning, and Operations*. Upper Saddle River, NJ: Prentice Hall.

Curtis, F.A. and K. Mak (1991), A medical waste management strategy, *Environmental management Health*, 2 (1), 13–18.

DeJohn, P. (2004), Ahead of the pack: Custom trays facing buyer scrutiny, *Hospital Materials Management*, 29 (4), 9–12.

Donston, D. (2007), Hospital finds recycling cure, *eWeek*, 24 (32), 22–26.

Environment Canada (2009), Pollution Prevention in the Health Sector, www.ec.gc.ca.

EPA (2005), *Profile of the Healthcare Industry*. Washington, DC: U.S. Environmental Protection Agency.

Fisher, B.E. (1996), Dissolving medical waste, *Environmental Health Perspectives*, 104 (7), 708–710.

Flinders Medical Centre (2000), Recycling; Segregation of Waste; Use of Recycled Products; Energy Conservation: Flinders Medical Centre, www.environmental-expert.com.

French, M.L. and R.L. LaForge (2006), Closed-loop supply chains in process industries: An empirical study of producer re-use issues, *Journal of Operations Management*, 24 (3), 271–286.

Geiselman, B. (2004), Waste awareness can save hospitals' costs, *Waste News*, 9 (28), 13.

Gilden, D.J., K.N. Scissors, and J.B. Reuler (1992), Disposable products in the hospital waste stream, *Western Journal of Medicine*, 156, 269–272.

Gilmore Hall, A. (2008), Greening healthcare: 21st century and beyond, *Frontiers of Health Services Management*, 25 (1), 37–43.

Hale, S. (2008), *Going Green without Sacrificing Quality*. Chicago, IL: Hospitals & Health Networks.

Hamilton, D.K. (2008), The challenge of sustainable hospital building, *Frontiers of Health Services Management*, 25 (1), 33–36.

Jang, Y.C., C. Lee, O.S. Yoon, and H. Kim (2006), Medical waste management in Korea, *Journal of Environmental Management*, 80 (2), 107–115.

Kaiser, B., P.D. Eagan, and H. Shaner (2001), Solution to health care waste: Life cycle thinking and 'green purchasing,' *Environmental Health Perspectives*, 1 (1), 7–12.

Kocabasoglu, C., C. Prahinski, and R.D. Klassen (2007), Linking forward and reverse supply chain investments: The role of business uncertainty, *Journal of Operations Management*, 25 (6), 1141–1160.

Kratz, K. and C. Thygesen (1992), A comparison of the accuracy of unit dose cart fill with the Baxter ATC-212 computerized system and manual filling, *Hospital Pharmacy*, 27, 19–21.

Landry, S. and M. Beaulieu (2007), The Hospital not just another link in the healthcare supply chain. In *Foundations of Production and Operations Management*, M.K. Starr (Ed.). Mason, OH: Thompson.

Landry, S. and R. Philipp (2004), How logistics can service healthcare, *Supply Chain Forum*, 5 (2), 24–30.

Messelbeck, J. and L. Sutherland (2000), Applying environmental product design to biomedical products research, *Environmental Health Perspective*, 108 (6), 997–1002.

Messelbeck, J. and M. Whalley (1999), Greening the health care supply chain: Triggers of change, models for success, *Corporate Environmental Strategy*, 6 (1), 38–45.

Rau, E.H., R.J. Alaimo, P.C. Ashbrook, S.M. Austin, N. Borenstein, M.R. Evans, H.M. French et al. (2000), Minimization and management of waste from biomedical research, *Environmental Health Perspectives*, 108 (6), 953–977.

Tudor, T.L., C.L. Noonan, and L.E.T. Jenkin (2005), Healthcare waste management: A case study from the national health service in Cornwall, United Kingdom, *Waste Management*, 25 (6), 606–615.

Weir, E. (2002), Hospitals and the environment, *Canadian Medical Association Journal*, 166 (3), 354.

INTERDISCIPLINARY IV RESEARCH ON CLOSED-LOOP SUPPLY CHAINS

VI

INTERDISCIPLINARY RESEARCH ON CLOSED-LOOP SUPPLY CHAINS

Chapter 12

Interdisciplinarity in Closed-Loop Supply Chain Management Research

Vishal Agrawal and L. Beril Toktay

Contents

12.1 Introduction

Closed-loop supply chain (CLSC) management has evolved to be a mature area of supply chain management in its own right. Guide and Van Wassenhove (2009) describe how this research began with focusing on tactical and operational issues such as the disassembly of products or shop-floor control and coordination, and later started addressing issues pertaining to the entire reverse supply chain such as product acquisition, supply-chain contracting, and incentives. Recently, the literature has begun studying issues at the intersection of other disciplines, drawing on concepts from industrial ecology, studying marketing issues such as the pricing and positioning of new and remanufactured products, etc. It is clear that the research in CLSCs is moving in an interdisciplinary direction. This is of great value, as the "messier" problems in industry tend to require an interdisciplinary approach.

At a recent workshop,* invited industry participants† shared the issues that they face in closing the loop in their business context. These questions were then grouped primarily along the areas of industrial ecology, new product development, marketing, and economic development, and discussed by small groups of workshop participants. This chapter builds on these discussions to present some existing research that is relevant to the questions posed, and is therefore necessarily interdisciplinary. A number of open research directions are also presented. We hope that describing industry problems and the current state of the art will prompt researchers to close existing gaps in the literature and increase the practical impact of CLSC management research. Table 12.1 provides

TABLE 12.1 List of Teaching Cases with Significant Closed-Loop Supply Chain Content and Interdisciplinary Issues

Product design and recovery	Vietor and Murray (1995)
Marketing of refurbished and remanufactured products	Van Wassenhove et al. (2002)
	Rayport and Vanthiel (2003)
Economic and environmental benefits of leasing	Olivia and Quinn (2003)
By-product synergy opportunities	Anderson and Mackenzie (2006)
	Lee et al. (2009)
Sustainability and supply chain management	Plambeck and Denend (2007)

* The 8th International Closed-Loop Supply Chain Workshop: Interdisciplinarity in Closed-Loop Supply Chain Management Research, October 9–11, 2008, College of Management, Georgia Institute of Technology. This workshop was partially funded by the National Science Foundation through Grant number DMI-0631954.
† John Wuichet, Installation Management Command, Southeast Region; Eric Nelson, Interface Americas; A.B. Short and Pat Robinson, MedShare International.

a list of teaching material that has significant CLSC content and that is at the interface of the above-mentioned areas, to aid in broadening instructions related to CLSC management.

This chapter is organized as follows. In Section 12.2, we provide short case studies based on each industry participant to provide an account of their CLSC activities and the problems they currently face in managing these operations. In Section 12.3, we identify the common threads running through the issues raised by these discussions and provide insights from the existing literature. We conclude this chapter in Section 12.4.

12.2 Case Studies

12.2.1 Interface, Inc.

Interface, Inc., is the world's largest provider of commercial carpet tile. It is well recognized for its attempts to be an environmentally sustainable enterprise with the ambitious goal of zero environmental impact by 2020 (Toktay et al. 2006). This is a particularly challenging goal, as the inputs into carpet production are primarily oil based, and recycling remains elusive, with 5 billion square yards of carpet deposited in U.S. landfills each year. Here, we focus on two interconnected elements in their road map to sustainability—closing the loop and redesigning commerce. The proof of concept at Interface would be the Evergreen Services Agreement (ESA), whereby carpet would be leased, not sold. With operating lease agreements, carpet would remain in the ownership of Interface and reclaimed carpet could be recycled, diverting it from the landfills. This practice of leasing, rather than selling, has been claimed to be an environmentally superior strategy for decreasing virgin material use, reducing waste generation, and closing the loop (Fishbein et al. 2000). The ownership of off-lease products would also provide the manufacturer with an incentive to utilize the residual value of these products. This is the rationale that prompted Interface to introduce ESA as part of their sustainability strategy as early as 1995. However, ESA was not successful. The main reason for its failure was the inability to recover value from used carpets as there was no viable technology to separate the face fiber from the backing, nor to recycle the face fiber consisting of Nylon 6,6 in the case of Interface; see Toktay et al. (2006, pp. 23–24) for a detailed discussion.

Interface's search for a way to close the loop was finally rewarded when they were introduced to an Italian company that manufactured machines to split leather to precise specifications (Ferguson et al. 2008). The technology was customized for the carpet industry to separate backing from face fiber, and to remelt the material back into pellets that could be fed back into the manufacturing process thanks to the very low levels of contamination achieved. Interface acquired this recycling process and introduced it as Re-Entry 2.0 in 2007. So far Interface has one facility in Georgia, but as this recycling process is a simple, low-footprint process, Interface is considering opening 19 such recycling facilities around the country to process 450 million lbs of reclaimed carpet per year (Nelson 2008).

At this juncture, some of the questions faced by Interface, Inc., are as follows:

1. How does Interface responsibly expand its carpet recycling operations? Should they use one centralized facility, build many smaller facilities across the United States, or license the technology? What are the environmental and economic implications of these choices?
2. Can Interface use recycling as a means to create economic development in local communities?
3. How should Interface's current manufacturing strategy be adapted in the face of the recycling opportunity?
4. Given that reclaimed carpet has some residual value now, is the reintroduction of the leasing program a good business strategy for Interface?

12.2.2 *Army Installation Management Command*

The army has initiated a sustainability planning process with a 25-year goal setting framework and a 5-year development plan overseen by the Army Installation Management Command (IMCOM). IMCOM's primary role is overseeing all facets of installation management including construction, barracks and housing, food management, etc. In this capacity, they handle the large variety and quantity of end-of-use and end-of-life streams generated in army installations. Below we discuss three projects developed by IMCOM-SE that aim to close the material loop and reduce the installations' environmental footprint.

IMCOM is working to close loops within local economies surrounding the installations by collaborating with external stakeholders. As such, the focus is on products for which there is manufacturing activity in the region. For example, furniture jobs are down 20 percent in North Carolina (where Fort Bragg is located) since 2007, while the on-site construction and demolition landfill cannot accept furniture, imposing a higher off-site disposal cost on the army. IMCOM promotes the design of furniture that is recyclable to achieve triple bottom-line benefits: obtaining value recovery from waste, reducing environmental impact, and generating jobs for local economies surrounding the installations. Some associated challenges are incorporating the "buy local" concept into IMCOM's centralized purchasing process and giving manufacturers enough incentive to design and build such products.

IMCOM has explored servicizing, and in particular, leasing, as an environmentally preferred alternative for such products like furniture and mattresses. Some of the challenges they face are finding an interested vendor, designing the right contract, and integrating into the purchasing-focused centralized procurement process.

IMCOM has also explored by-product synergies, where waste from one operation can be used as a raw material for another operation. Most installations occupy forested land that generates large quantities of timber harvest waste. Another large source of wood waste is demolition debris. Historically, timber waste was burned on the installations and demolition waste was landfilled, while bark fuel was purchased to run a central energy plant. Some installations now utilize their wood waste as fuel for the energy plant, which reduces disposal cost, fuel procurement cost, and the number of smoke days. There could be several such synergies in IMCOM's operations, but the challenge lies in identifying and managing such symbiotic activities.

In this context, some of the questions faced by IMCOM are the following:

1. How can the command encourage the design and availability of greener products?
2. The army procurement systems are becoming more centralized, but the disposal and recycling activities are carried out locally. Should closed-loop systems be pursued at the local or the global scale?
3. How can closing the loop best be leveraged to generate local employment opportunities?
4. What processes should be deployed to generate and manage by-product synergies?
5. Would leasing help to reduce costs and avoid end-of-life management? If so, should the command lease directly from a manufacturer or through a third-party lessor? How should existing internal processes be redesigned to support such an endeavor?

12.2.3 MedShare International

Usable medical supplies worth $6.5 billion are discarded in the United States every year. Medshare International, an Atlanta-based nonprofit humanitarian organization, collects unused medical supplies in the United States and distributes them to needy facilities in other parts of the world. While the main driver of MedShare is their humanitarian impact, as they divert waste from U.S. landfills, their operations also lend themselves to claims of environmental benefits. The ability to legitimately claim that environmental considerations are included in MedShare's recipient selection decisions has important ramifications for their business: As MedShare depends on the philanthropic donations of individuals and foundations, and foundations have strict guidelines on what types of projects are in line with their charitable objectives, the ability to demonstrate a sensitivity to the environmental impact of their operations can significantly increase the number of foundations to which MedShare can submit proposals. Thus, MedShare is interested in expanding their recipient strategy to include both humanitarian and environmental objectives, and to have a systematic procedure for doing so. Issues to take into account are the environmental impact of the additional transportation created by their operations, the usage rate of donated supplies, and the quality of disposal methods in the recipient country (Denizel et al. 2009).

As MedShare operates on donations from sponsors, they are interested in figuring out whether they can be more effective by improving the cost efficiency of their collection operations (without increasing their environmental impact). One option is to partner with waste collectors, because they visit the same hospital facilities as MedShare. In addition, Medshare could integrate another facility or partner to re-sterilize the collected supplies that would otherwise have to be discarded. However, these partnerships may increase the operational risks and constraints for MedShare.

MedShare currently has two locations in the United States. One is on the east coast, in Atlanta, and the other is on the west coast, near San Francisco. Transporting supplies to these two central facilities is a major expense and contributes to their carbon footprint. On the other hand, there are economies of scale associated with a central facility. In addition, as shipments to each recipient are based on inventory at only one facility, centralizing the inventory provides a broader array of products for the recipients to choose from. The question for MedShare is whether to utilize a distributed or centralized approach as they continue to grow.

In summary, some of the questions faced by MedShare International are as follows:

1. How should MedShare measure the relative environmental impact of different potential recipients?
2. How can MedShare balance its dual objectives of humanitarian and environmental benefits?
3. Would partnering with medical waste collectors or other organizations improve the economic and environmental performance of their collection process?
4. How should Medshare expand to other communities or countries?

12.3 Relevant Interdisciplinary Research

The above case studies highlight that although each one of these organizations is very different, the main issues they are grappling with have common threads at the interfaces with the following research areas: industrial ecology and supply chain management, identifying and managing by-product synergies, designing and implementing product–service systems, developing and marketing greener products, and creating economic development opportunities by closing the loop. In this section, we will provide some insights on questions including, but not limited to, the specific ones discussed in the case studies. Our goal is not to provide an exhaustive review of the literature pertaining to each question, but to highlight different interdisciplinary approaches to such questions through a limited, but representative subset of the literature.

12.3.1 Industrial Ecology and Supply Chain Management

The main tenet of industrial ecology is to take a systems view and consider holistic analyses that focus on multiple attributes of a system (e.g., different dimensions of environmental impact). Due to this perspective, research in the industrial ecology domain considers the environmental impact through the entire product life cycle. As the product goes through the different stages of the supply chain, it also goes through different phases of its life cycle; because of this, existing approaches in industrial ecology and CLSC management are complementary in nature. This synergy is reflected in recent interdisciplinary researches that address environmental issues using tools from both domains. For example, Faruk et al. (2002) describe a framework that firms can use to assess the environmental impact of their entire supply chain. Rosen et al. (2001) discuss different supply-chain contracting mechanisms that can be used by firms in the computer industry to incentivize their suppliers to improve the environmental quality of their products. Rock et al. (2006) conduct a case study of Motorola's global supply chain to analyze if the internal environmental standards are adopted by its suppliers in other countries. In this section, we will focus exclusively on literature that draws on both domains and considers both the profitability and the environmental performance of CLSCs.

Subramanian et al. (2008) develop a decision support tool using a mathematical programming model for a profit-maximizing manufacturer to capture environmental considerations such as sustainable product design, management of emission allowances, and loop-closing activities such as recovery, remanufacturing, and disposal along with traditional operational considerations such as capacity, production, and inventory.

Matthews et al. (2002) assess the economic and environmental implications of the centralization of inventory and warehousing. Although increased centralization reduces the inventory level and the number of warehouses, and consequently reduces inventory and warehousing costs and the environmental impact of warehousing, it increases the costs and environmental impact of transportation. Based on a case study of the spare-parts inventory at U.S. Department of Defense warehouses, they found that since spare-parts have low demand, there are both significant economic and environmental benefits from centralization. Although these insights may only hold for low-demand products, their analysis provides a framework that can be used to investigate any product category and various design and location decisions arising in the management of CLSCs.

Quariguasi Frota Neto and Bloemhof (2008) examine the economic and environmental implications of closing the loop by recovering, remanufacturing, and selling computers and mobile phones. They use the concept of eco-efficiency, which is defined as economic output per unit of environmental impact. As consumers discount remanufactured products, the remanufactured product price (economic output) is generally lower than the new product price. Remanufactured products are nevertheless more eco-efficient (with energy consumption as the environmental impact measure), because the energy consumption associated with remanufacturing is far lower than that associated with producing new products.

Geyer and Jackson (2004) study "supply loops": end-of-life strategies that divert waste and replace primary raw materials in forward supply chains. Using examples from the construction industry, they argue that such supply loops can not only result in environmental benefits but also help firms reap economic benefits. However, firms need to take into account that a component may be reused only a finite number of times due to technical constraints and enjoy market demand only for a limited time. The main insight for firms is that they need to align recovery decisions such as collection and remanufacturing rates with such product characteristics (Geyer et al. 2007).

These frameworks and insights are valuable to entities like Interface and Medshare, who are interested in incorporating both cost and environmental considerations in designing their CLSCs. Nevertheless, such applications are not common, and there is an opportunity to both do proof-of-concept on industry problems and to develop frameworks that are specific to reverse (rather than forward) supply chains.

12.3.2 By-Product Synergy and Industrial Symbiosis

By-product synergy is a process by which wastes can be converted into marketable commodities, and industrial symbiosis is the exchange of wastes, by-products, and different forms of waste energy among closely situated firms in an industrial complex. In their seminal paper, Ehrenfeld and Gertler (1997) describe the benefits and challenges involved in managing opportunities for by-product synergy and industrial symbiosis using the example of the industrial district of Kalundborg, Denmark. Although the benefits of by-product synergy and industrial symbiosis are well recognized, there are several managerial challenges associated with benefiting from them (see Anderson and Mackenzie 2006 and Lee et al. 2009 for related teaching cases). The initial challenge is the identification of such opportunities: As they are not the main focus of the business, employees have no incentive to identify or champion such causes. Even if such opportunities are identified, finding reliable markets for by-products may be difficult. As they require long-term commitment and there is considerable uncertainty in their success, investing and implementing in such opportunities is also difficult.

There is little research literature on how a business can maximize the economic benefits from by-product synergy and industrial symbiosis. Lee (2009) considers a firm operating in a competitive setting, where it can convert a waste stream into a marketable by-product. If the sale of the by-product is profitable enough, it may incentivize the firm to generate more waste, which will promote greater production of the base product and increase consumption. Thus, the net environmental impact of such opportunities could potentially be negative and a firm needs to be careful about claiming an environmental benefit when it implements such by-product synergies. Interesting avenues to explore are how to price by-products when there is uncertainty in both supply and demand for them, and how to incentivize employees to identify and manage by-product synergies.

12.3.3 Product–Service Systems

A product–service system is defined as a product and a service combined in a system to deliver consumer needs and reduce environmental impact, typically by displacing new production or increasing usage efficiency (Baines et al. 2007). Primary examples of product–service systems are servicizing, renting, leasing, sharing, or pooling through membership schemes. Offering maintenance, take-back, or disposal services are also considered to be part of product service systems. In this chapter, we focus on these systems from the point of view of a profit-maximizing firm. A recurring theme in this stream of literature is that such product–service systems do not necessarily yield superior environmental outcomes. The interested reader is directed to Mont (2004), who provides a detailed discussion of other aspects of these systems.

An interesting example of servicizing is for the consumption of indirect materials such as solvents or hazardous catalysts in the chemical industry and to a lesser extent in the electronics and automotive industries. A buyer would like to reduce his consumption of such indirect materials, but the supplier has an incentive to sell a larger volume and will not invest in reducing the consumption. Using examples from industrial practice, Reiskin et al. (2000) describe how the traditional supplier–customer price-based relationship can be transformed to where the supplier does not sell, but provides these indirect materials as a service. This might provide the supplier with an incentive to reduce consumption. Corbett and DeCroix (2001) study contracting schemes for sharing savings from such servicizing opportunities in a supply chain, where both the suppliers and the customers can benefit from dematerialization. They show that such contracts can increase the supply chain profits but at the expense of increased consumption, which leads to environmentally inferior outcomes.

Leasing is a strategy that has long been used with the goal of maximizing firm profits. Recently, the industrial ecology literature has promoted leasing as environmentally superior to selling (Hawken et al. 1999, Fishbein et al. 2000, Lifset and Lindhqvist 2000, Robert et al. 2002). The rationale is that as the firm maintains ownership of the off-lease products, it has an incentive to refurbish and remarket the product, which helps extend its useful life, divert waste from landfills, and displace new production. However, some argue that if manufacturers have control over off-lease products, they will prematurely dispose them off to reduce cannibalization and lead to environmentally worse outcomes (Lawn 2001).

Agrawal et al. (2009b) compare leasing and selling from the manufacturer's point of view to identify conditions under which leasing can be both financially and environmentally superior from a life-cycle perspective. They find that manufacturers would find it profitable to lease only if they face a lower disposal cost than the consumers. Commercial carpeting, while durable, does not lend itself to reuse and only some of the material can be recycled. Thus, by committing itself to collecting and (partially) recycling the carpet, Interface effectively increased its disposal cost significantly relative to local landfilling by its customers. Consequently, it is not surprising that despite originally being championed at the highest levels of the company, the leasing program was phased out.

Agrawal et al. (2009b) also find that even if a leasing firm does not prematurely dispose off-lease products, it may still have an incentive to produce a larger quantity of products, which negates any reduction in the disposal impact. The main message for firms considering promoting or adopting leasing programs for improving their environmental performance is to carefully consider the disposal cost scenarios, the product durability, and the environmental impact over the entire product life cycle.

Membership or pooling schemes such as car-sharing services can be environmentally beneficial as they may induce consumers to participate in these schemes instead of buying new products, thereby leading to lower consumption and production. However, Bellos et al. (2009) study membership schemes in the context of transportation and show that membership schemes may lead to inferior environmental outcomes where they induce consumers who would otherwise use lower-impact substitutes such as public transport to join such schemes.

Olivia and Quinn (2003) is a case study based on Interface's Evergreen Leasing program that highlights the managerial challenges associated with its implementation. The main insight from the literature discussed above is that a firm implementing such a product–service system needs to be careful before claiming environmental superiority. This is of importance to Interface in the context of the revival of the Evergreen Leasing Program and to IMCOM in its efforts to promote leasing as an environmentally superior procurement strategy.

There are several open questions regarding such systems. One direction for future research is whether an original equipment manufacturer (OEM) should lease directly to consumers or sell to a third-party lessor who would lease to consumers. Another question is how to manage conflicts with existing dealer networks while introducing such new systems. Can the inclusion of ancillary services such as maintenance help increase the attractiveness of such options for the consumers? Finally, more research is needed to investigate the appropriate design of lease terms to achieve both economic and environmental benefits.

12.3.4 New Product Development

The literature in new product development and innovation has studied problems such as quality choice, product line design, and component commonality. These issues are relevant in the CLSC context, albeit with complementary considerations such as the role of green consumers and the effect of take-back legislation. Here, we focus on the managerial challenges associated with new product development in the CLSC context, most of which are nicely captured in Vietor and Murray

(1995), a case study describing Xerox's attempts to align its design strategy with its recovery strategy. A discussion of the engineering design aspects is the subject of Chapter 4.

Chen (2001) studies the design problem for a firm where the conventional and environmental attributes of a product conflict with each other and the market may consist of traditional and green consumers. They show that a firm's design problem critically depends on the legislative (presence of regulation) and market conditions (population of green consumers). They show that in the presence of government policies such as stricter environmental standards, the firm's design and marketing strategies may change, resulting in inferior environmental outcomes. Subramanian et al. (2009) analyze a firm's component commonality decision in the presence of recovery and remanufacturing operations. They show that ignoring remanufacturing operations at the product design stage can have a detrimental impact on a firm's profitability. This stream of literature emphasizes the importance of product design in leveraging the benefits from closing the loop.

An important consideration in the design and introduction of new products is the presence of take-back legislation (see Chapter 3 for an in-depth discussion of this type of legislation). Plambeck and Wang (2009) study the impact of e-waste legislation on new product introduction and find that "fee-upon-sale" type of legislation decreases the rate of new product introduction, and consequently, the quantity of e-waste, but does not provide firms the incentive to design products for recyclability. They also find that e-waste legislation that imposes a "fee-upon-disposal" does not reduce the rate of introduction and e-waste, but leads to firms designing products for recyclability. Atasu and Subramanian (2009) analyze the effect of legislation on designing products for recyclability. They find that individual producer responsibility programs offer higher incentives for recyclable product design as compared to collective responsibility programs.

In the absence of legislation, the benefits from designing for the environment depend on the presence of green consumers. Ginsberg and Bloom (2004) discuss different types of consumers and their preferences regarding green products. They say that somewhere around 15–46 percent of consumers are interested in some form of green product. However, only a very small fraction (at most 5–10 percent) of these consumers would spend more to buy a "greener" product. Moreover, not all industries or products enjoy an already existent consumer population who are willing to pay a premium for greener products. In such a setting, the question is how a firm or a policy-maker can encourage the growth of such consumer segments to support the development of environmentally superior products.

Andrews and DeVault (2009) use a multi-heterogeneous-agent simulation to analyze the interactions between firm strategies, government policy, and consumer preferences and study the emergence of green markets using an application to hybrid cars. Their insights are useful for different stakeholders: Firms can innovate to create greener products either as a response to or in anticipation of government regulation. However, green markets will not emerge unless there are enough green consumers. Thus, innovation by firms is necessary, but not sufficient; the presence of green consumers is also important. They find that governmental intervention such as bans on environmentally inferior products or taxes only help preserve niche-green markets. The only way for such markets to grow and for green products to go mainstream is through cost parity. This implies that to increase the availability of recyclable products, entities such as IMCOM may benefit from joining groups such as the "Buy Recycled Business Alliance," which would help to increase demand for recycled products and help achieve economies of scale leading to cost parity with ordinary products.

There are several open questions for future research. One such question is whether it is profitable to design products for modularity so that it is easier to reuse them and maximize value recovery. Another is the trade-off between the ability to innovate and the ability to benefit from returned cores in subsequent generations.

12.3.5 Marketing

The profitability of CLSCs depends on the market acceptance of recyclable, refurbished, or remanufactured products. Marketing such products poses several challenges for a firm. Remanufactured products may potentially cannibalize the demand for the firm's new products. Thus, the joint positioning and pricing of new and remanufactured products is a key problem faced by the firm. We direct the interested reader to Chapter 2 for a detailed discussion of the trade-offs involved. In this section, we will focus on the literature that studies the effect of consumer perceptions of remanufactured and refurbished products on a firm's closed-loop strategies.

Consumers may have quality and reliability concerns regarding remanufactured products, which may lead to a lower perceived value, which may in turn inhibit their market acceptance or profitability. This has extensively been used as a modeling assumption in the literature (Debo et al. 2005, 2006, Ferguson and Toktay 2006, Atasu et al. 2008). Recent research efforts have validated this assumption through experimental and empirical analyses. Guide and Li (2009) conduct eBay experiments using a consumer product (a power tool) and a commercial product (internet router) and find that on average, remanufactured products are purchased at lower prices than new products. They also find little overlap between bidders for the consumer product, but greater overlap for the commercial product. Thus, a commercial product firm should be more careful about the cannibalization of its new products by its remanufactured products. Subramanian and Subramanyam (2009) use purchase data from eBay and show that the price differential depends on the seller reputation and the product category (see Chapter 8 for more details). Agrawal et al. (2009a) conduct an experiment using Apple iPods and find that the subjects have a lower willingness to pay for remanufactured products and that they have a higher perceived value for an OEM-remanufactured product as compared to a third-party–remanufactured product.

The presence of remanufactured products may raise quality issues concerning the firm's new products: "… Just where are the refurbished iPhones coming from? Is Apple getting enough returns so they can resell them …" (CNET 2007). Existing literature in marketing has established that consumers' perceptions spillover between different products sold under the same brand (Sullivan 1990, Rangaswamy et al. 1993, Erdem 1998). As remanufactured products are functionally and physically the same as the new product, one may expect consumer perception of the new product changing in the presence of its remanufactured counterpart. Indeed, Agrawal et al. (2009a) establish that the presence of remanufactured products has a significant impact on the consumers' perceived value for new products. They find that the presence of OEM-remanufactured products lowers the value of new products, but the presence of third-party–remanufactured products increases the value of new products. This result implies that while selling remanufactured products, an OEM should alleviate consumer concerns regarding quality or reliability by providing more information regarding the remanufacturing processes or better warranties. The authors also show that when such consumer perceptions are taken into account, an OEM's optimal remanufacturing and preemption strategy may drastically change. The main insight from this stream of literature is that firms should first investigate the consumer perceptions for their product and manage their CLSCs accordingly.

Recently, research has also focused on different information cues that can help to increase the perceived value of remanufactured products. Ovchinnikov (2009) conducts an experiment using Dell laptops where consumers are provided with the price differential between the new and remanufactured products as a cue and studies their valuation for the remanufactured product. He finds that the fraction of consumers who switch from a new to a remanufactured product first increases and then decreases (inverted-U shape) as the discount on the remanufactured products increases. This provides some evidence that consumers may infer the quality of the remanufactured product based on the price. Quariguasi Frota Neto (2008) explicitly considers the role of the new-product price as a reference price and finds that using the new product price helps increase the consumers' willingness to pay for remanufactured products. Agrawal et al. (2009a) conduct an experiment to examine the role of information regarding the availability of remanufactured products on the consumers' perceived value for new and remanufactured products. They find that the perceived value of remanufactured products decreases with an increase in their availability. Thus, the firm may benefit from restricting information regarding the availability of remanufactured products. The main insight from this literature is that a firm can improve the acceptance and value of recovered products by using the appropriate marketing cues and strategies.

There are other marketing factors and cues that warrant further investigation. For example, while it has been recognized that the reputation of third-party remanufacturers has a significant impact on the perceived value of remanufactured products (Subramanian and Subramanyam 2009), the effect of OEM reputation on consumer perceptions is still unknown. More research is also needed on analyzing the impact of other factors on consumer perception, such as warranties and information regarding the source of cores used for remanufacturing.

12.3.6 Economic Development

A testament to the impact of closing the loop on economic development is the contribution of the remanufacturing sector to the U.S. economy. According to a 1996 survey conducted in the United States, 79 different product areas were being remanufactured, employing 480,000 people and consisting of 73,000 different firms, which is comparable to other mainstream industries (see Lund 2001). Goldman and Ogishi (2001) argue that new activities from the diversion and reuse of waste can result in economic development in economically distressed areas. Smith and Keoleian (2004) use a life-cycle assessment model to analyze the environmental benefits of remanufacturing automotive engines. They also show that remanufacturing can also have social benefits either through additional employment opportunities or through greater affordability of the products for small businesses and consumers.

Leigh and Patterson (2005) discuss how recycling construction debris can not only result in environmental benefits but also assist in economic development as such activities result in the creation of jobs for low-income, low-skilled residents. Leigh and Realff (2003) study the economic development potential of the recycling and reuse of computers in the state of Georgia. An interesting question they pose is whether end-of-life material flows can be designed to both promote economic development in the severely distressed areas of Atlanta and limit the environmental performance of collection. Using census and demographic data, they estimate the quantity of obsolete computers in households across the state. Taking existing electronics stores as collection centers, they compare two locations for placing a recycling network along the economic development and environmental dimensions. The "greenfield" location is a traditionally affluent section of the city, where most of the e-waste originates and which is closer to major transport routes, but is further away from sources of the low-skilled unemployed labor force. In contrast, the "greyfield" location is a traditionally industrial location that is further away from sources of e-waste but closer to the sources of required labor. They find that locating the recycling center in the greyfield location has lower environmental impact and also results in more economic development by being closer to the unemployed labor force.

The nexus of economic development and CLSCs has not been studied except for the above-mentioned papers and is a fertile area for combining the social and environmental benefits of CLSCs. Building robust statistical models for estimating the volume of used products that can be collected based on demographic statistics and on consumer behavior regarding recycling activities would help evaluate the potential for employment creation. Developing frameworks to incorporate information about characteristics of the workforce in supply chain design decisions would also be valuable.

12.4 Conclusions

We hope that the case studies presented have highlighted the interdisciplinary nature of issues that closing the material loop raises. As evidenced by the (mostly recent) research using approaches from different disciplines, there is a growing recognition that challenges faced by industry to design and manage CLSCs cannot be solved by using a single approach or by only drawing from the knowledge base of one domain. Indeed, as discussed in this chapter, there are many opportunities that remain for us to collaborate with researchers in other disciplines and help to increase the influence of our research on managerial practice.

Acknowledgments

The case descriptions were primarily based on presentations by Eric Nelson (Interface), John Wuichet (J M Waller Associates for IMCOM-SE), A.B. Short (MedShare), and Pat Robinson (MedShare). We gratefully acknowledge the valuable input of all workshop participants. This work was funded by NSF Grants DMI-0631954 and DMI-0522557.

References

Agrawal, V., A. Atasu, and K. van Ittersum. 2009a. The Effect of Consumer Perceptions on Competitive Remanufacturing Strategies. *Working Paper*. College of Management, Georgia Institute of Technology, Atlanta, GA.

Agrawal, V., M. Ferguson, L. B. Toktay, and V. Thomas. 2009b. Is Leasing Greener Than Selling? *Working Paper*. College of Management, Georgia Institute of Technology, Atlanta, GA.

Anderson, T. and S. Mackenzie. 2006. Applied Sustainability LLC: Making a Business Case for By-Product Synergy. Stanford Graduate School of Business, Case E-118, Stanford, CA.

Andrews, C. and D. DeVault. 2009. A model with heterogeneous agents. *J. Ind. Ecol.* 13(2) 326–345.

Atasu, A. and R. Subramanian. 2009. The Effect of Extended Producer Responsibility on Recyclable Product Design. *Working Paper*. College of Management, Georgia Institute of Technology, Atlanta, GA.

Atasu, A., M. Sarvary, and L. N. Van Wassenhove. 2008. Remanufacturing as a marketing strategy. *Manage. Sci.* 54(10) 1731–1747.

Baines, T., H. Lightfoot, S. Evans, A. Neely et al. 2007. State-of-the-art in product-service systems. *Proc. Inst. Mech. Eng. Part B: J. Eng. Manuf.* 221(10) 1543–1552.

Bellos, I., M. Ferguson, and L. B. Toktay. 2009. To Sell or to Provide? A Comparison of Selling and Membership. *Working Paper.* College of Management, Georgia Institute of Technology, Atlanta, GA.

Chen, C. 2001. Design for the environment: A quality-based model for green product development. *Manage. Sci.* 47(2) 250–263.

CNET. 2007. Apple Sells Refurbished iPhones. http://news.cnet.com/8301-17938_105-9762502-1.html.

Corbett, C. and G. DeCroix. 2001. Shared-savings contracts for indirect materials in supply chains: Channel profits and environmental impacts. *Manage. Sci.* 47 881–893.

Debo, L. G., L. B. Toktay, and L. N. Van Wassenhove. 2005. Market segmentation and product technology selection for remanufacturable products. *Manage. Sci.* 51(8) 1193–1205.

Debo, L. G., L. B. Toktay, and L. N. Van Wassenhove. 2006. Life-cycle dynamics for portfolios with remanufactured products. *Prod. Oper. Manage.* 15(4) 498–513.

Denizel, M., M. Ferguson, and L. B. Toktay. 2009. Multiple-Criteria Decision Models for Recipient Selection in Donations of Medical Supplies. *Working Paper.* College of Management, Georgia Institute of Technology, Atlanta, GA.

Ehrenfeld, J. and N. Gertler. 1997. Industrial ecology in practice: The evolution of interdependence at Kalundborg. *J. Ind. Ecol.* 1(1) 67–79.

Erdem, T. 1998. An empirical analysis of umbrella branding. *J. Mark. Res.* 35 339–351.

Faruk, A., R. Lamming, P. Cousins, and F. Bowen. 2002. Analyzing, mapping and managing environmental impacts along supply chains. *J. Ind. Ecol.* 5(2) 13–36.

Ferguson, M. E. and L. B. Toktay. 2006. Manufacturer strategies in response to remanufacturing competition. *Prod. Oper. Manage.* 15(3) 351–368.

Ferguson, M., E. Plambeck, and L. Denend. 2008. *Teaching Note for Interface's Evergreen Services Agreement.* Harvard Business Publishing, Harvard University, Cambridge, MA.

Fishbein, B., L. McGarry, and P. Dillon. 2000. Leasing: A Step towards Producer Responsibility. Technical Report, INFORM, Inc., New York.

Geyer, R. and T. Jackson. 2004. Supply loops and their constraints: The industrial ecology of recycling and reuse. *Calif. Manage. Rev.* 46(2) 55–73.

Geyer, R., L. Van Wassenhove, and A. Atasu. 2007. The economics of remanufacturing under limited component durability and finite product life cycles. *Manage. Sci.* 53(1) 88.

Ginsberg, J. and P. Bloom. 2004. *Choosing the Right Green Marketing Strategy. MIT Sloan Management Review*, MIT, Cambridge, MA.

Goldman, G. and A. Ogishi. 2001. The Economic Impact of Waste Disposal and Diversion in California. Technical Report to the California Integrated Waste Management Board, University of California, Berkeley, CA.

Guide, D. and K. Li. 2009. The Potential for Cannibalization of New Product Sales by Remanufactured Products. *Working Paper.* Smeal College of Business, The Pennsylvania State University, University Park, PA.

Guide, D. and L. N. Van Wassenhove. 2009. OR FORUM-the evolution of closed-loop supply chain research. *Oper. Res.* 57(1) 10–18.

Hawken, P., A. Lovins, and L. Lovins. 1999. *Natural Capitalism.* Little, Brown & Company, New York.

Lawn, P. 2001. Goods and services and the dematerialization fallacy: Implications for sustainable development indicators and policy. *Int. J. Serv. Technol. Manage.* 2(3–4) 363–376.

Lee, D. 2009. Turning Waste into By-Product. *Working Paper*. Harvard Business School, Harvard University, Boston, MA

Lee, D., M. Toffel, and R. Gordon. 2009. Cook Composites and Polymers Co. Harvard Business School Case 9-608-055, Harvard University, Cambridge, MA.

Leigh, N. and L. Patterson. 2005. Construction and Demolition Debris Recycling for Environmental Protection and Economic Development. Technical Report, Southeast Environmental Finance Center, University of North Carolina at Chapel Hill, NC.

Leigh, N. and M. Realff. 2003. A framework for geographically sensitive and efficient recycling networks. *J. Environ. Plann. Manage.* 46(2) 147–165.

Lifset, R. and T. Lindhqvist. 2000. Does leasing improve end of product life management? *J. Ind. Ecol.* 3(4) 10–13.

Lund, R. 2001. Remanufacturing as a resource. In *Fifth International Congress on Environmentally Conscious Design and Manufacturing*, June 16–17, Rochester, NY.

Matthews, H., C. Hendrickson, and L. Lave. 2002. The economic and environmental implications of centralized stock keeping. *J. Ind. Ecol.* 6(2) 71–81.

Mont, O. 2004. *Product-Service Systems: A Panacea or Myth?* The International Institute for Industrial Environmental Economics, Lund University, Lund, Sweden.

Nelson, E. 2008. The business of old carpet: The interface journey to 2020. In *The International Closed-Loop Supply Chain Workshop: Interdisciplinarity in Closed-Loop Supply Chain Management Research*, October 9–11, 2008, College of Management, Georgia Institute of Technology, Atlanta, GA.

Olivia, R. and J. Quinn. 2003. Interface's Evergreen Services Agreement. Harvard Business School Case 9-603-112, Harvard University, Cambridge, MA.

Ovchinnikov, A. 2009. Revenue and Cost Management for Remanufactured Products. *Working Paper*. Darden Graduate School of Business, University of Virginia, Charlottesville, VA.

Plambeck, E. and L. Denend. 2007. Wal-Mart's Sustainability Strategy. Stanford Graduate School of Business, Case OIT-71, Stanford, CA.

Plambeck, E. and Q. Wang. 2009. Effects of e-waste regulation on new product introduction. *Manage. Sci.* 55(3) 333–347.

Quariguasi Frota Neto, J. 2008. Eco-efficient supply chains for electrical and electronic products. Dissertation, Rotterdam School of Management, Erasmus University, Rotterdam, the Netherlands.

Quariguasi Frota Neto, J. and J. Bloemhof. 2008. The Environmental Gains of Remanufacturing: Evidence from the Computer and Mobile Industry. *Working Paper*. Rotterdam School of Management, Erasmus University, Rotterdam, the Netherlands.

Rangaswamy, B., R. Burke, and T. Oliva. 1993. Brand equity and extendibility of brand names. *Int. J. Res. Mark.* 10 61–75.

Rayport, J. and J. Vanthiel. 2003. Green Marketing at Rank Xerox. Harvard Business School Case 9-594-047, Harvard University, Cambridge, MA.

Reiskin, E., A. White, J. Johnson, and T. Vorta. 2000. Servicizing the chemical supply chain. *J. Ind. Ecol.* 3(2–3) 19–31.

Robert, K. H., B. Schmidt-Bleek, J. Aloise de Larderel, G. Basile, J.L. Jansen, R. Kuehr, P. Price Thomas, M. Suzuki, P. Hawken, and M Wackernagel. 2002. Strategic sustainable development—selection, design and synergies of applied tools. *J. Cleaner Prod.* 10(3) 197–214.

Rock, M. T., P. L. Lim, and D. P. Angel. 2006. Impact of firm-based environmental standards on subsidiaries and their suppliers: Evidence from Motorola-Penang. *J. Ind. Ecol.* 10(1–2) 257–278.

Rosen, C., J. Bercovitz, and S. Beckman. 2001. Environmental supply-chain management in the computer industry. *J. Ind. Ecol.* 4(4) 83–103.

Smith, V. and G. Keoleian. 2004. The value of remanufactured engines. *J. Ind. Ecol.* 8(1–2) 193–221.

Subramanian, R. and R. Subramanyam. 2009. Key Drivers in the Market for Remanufactured Products: Empirical Evidence from eBay. *Working Paper.* College of Management, Georgia Institute of Technology, Atlanta, GA.

Subramanian, R., B. Talbot, and S. Gupta. 2008. An Approach to Integrating Environmental Considerations within Managerial Decision-Making. *Working Paper.* College of Management, GeorgiaInstitute of Technology, Atlanta, GA.

Subramanian, R., M. Ferguson, and L. B. Toktay. 2009. The Impact of Remanufacturing on the Component Commonality Decision. *Working Paper.* College of Management, Georgia Institute of Technology, Atlanta, GA.

Sullivan, M. 1990. Measuring image spillovers in umbrella-branded products. *J. Bus.* 63(3) 309–329.

Toktay, L. B., L. Selhat, and R. Anderson. 2006. Doing well by doing good: Interface's vision of being the first industrial company in the world to attain sustainability. In *Enterprise Transformation: Understanding and Enabling Fundamental Change*, W. Rouse (Ed.), Wiley, New York.

Van Wassenhove, L., D. Guide, and N. Kumar. 2002. Managing Product Returns at Hewlett Packard. INSEAD Case 602-039-1.

Vietor, R. and F. Murray. 1995. Xerox: Design for the Environment. Harvard Business School Case 9-794-022, Harvard University, Cambridge, MA.

Chapter 13

Empirical Studies in Closed-Loop Supply Chains: Can We Source a Greener Mousetrap?

Stephan Vachon and Robert D. Klassen

Contents

13.1 Introduction

Since the 1990s, two major trends have affected manufacturing organizations around the world. First, it almost goes without saying that international competition has increased dramatically with globalization and multilateral trade accords. The increased competition has allowed customers to demand lower prices, better quality, and faster delivery, and at the industry level, it has pushed toward consolidation. In other words, many manufacturing supply chains have become both more efficient and more effective.

Multiple stakeholders, such as customers, consumers, regulators, and nongovernmental organizations, have fueled the second trend that is central to this chapter. These stakeholder groups have begun to expect, to varying degrees, better environmental stewardship from manufacturing firms and their supply chains (Delmas and Toffel 2008; Sharma and Henriques 2005). This can take on many forms, including end-of-life product responsibility, reduced wastes, and lower consumption of natural resources and energy. On the surface, these two trends might be perceived as incompatible, because moving waste products back through the supply chain increases logistical costs and inventory levels. Yet, a number of studies have offered evidence that environmental management is positively linked to operational and financial performance (King and Lenox 2002; Klassen and McLaughlin 1996; Vachon and Klassen 2006b; Zhu and Sarkis 2004).

Although popular environmental management practices such as reduction, reuse, and recycling (3Rs) are increasingly used within internal operations, a growing number of manufacturing organizations are beginning to explore how these practices might be extended outside their boundaries to taking back nearly new or end-of-life products. In the case of products recently sold but then returned by dissatisfied customers (i.e., early life returns), there is the expectation of friendly product return policies (Autry et al. 2001). At the other extreme, regulations (and to a lesser extent, customers) are also expecting firms to take responsibility for their products at the end of their useful life (i.e., end-of-life returns). Chapter 3 summarizes most of the environmental legislation and regulations regarding extended producer responsibility. Finally, packaging and other containers remain a challenge for consumable products because of their relatively low value and density (i.e., packaging returns) (Gonzalez-Tore et al. 2004).

In a few industries, such practices are not new. For example, major soft drink and beer companies in many regions have developed systems to recover bottles and cans for reuse or recycle (Stock 1998). However, with more sophisticated goods or a greater geographic diffusion of business customers or consumers, applying 3R principles through the supply chain to bring back products from downstream entities has proven to be difficult (Guide and Van Wassenhove 2002; Guide et al. 2000; Vachon et al. 2001). Overcoming these challenges is essential, both to retain value already embedded in materials and components and to divert waste away from landfills. However, to be competitively sound, manufacturing firms must also develop business models that integrate strategic and tactical decisions when implementing a closed-loop supply chain (CLSC).

This chapter has two main objectives: to overview the prior work and to explore new directions. First, the empirical research literature is reviewed with a particular focus on studies that have used survey and archival data to identify important constructs and relationships. Although our focus is on these two research methodologies, we should stress that other empirical methods, such as case studies, are important too, as many industry-, supply chain-, and firm-level case studies have been used to illustrate a conceptual model, or parameterize an analytic model. Based on our review of some key concepts and findings, we highlight some existing gaps and provide some new research questions. The second objective is to propose a conceptual model that addresses a number of the research questions or gaps found in the literature. These two objectives are addressed, respectively, in Sections 13.2 and 13.3. In Section 13.4, we present some future research avenues.

13.2 Overview of Literature

Although the terms and definitions used to describe various aspects of CLSC vary widely and tend to be somewhat imprecise in the research literature and managerial articles, it is helpful to briefly delineate a few terms, such as reverse logistics, reverse supply chain, and CLSC (French and LaForge 2006). These terms encompass gradually larger systems of materials, tasks, information, and strategy. Reverse logistics focuses on the storage and movement of materials (and associated information) from the consumer or end user back to the manufacturer, recycler, or other third party. The reverse supply chain integrates activities and interactions between parties in the supply chain for either early-use or end-of-life product flows, including tasks such as separation, reuse, remanufacturing, and recycling. Finally, CLSC includes the management of both the forward (or conventional) supply chain and the reverse supply chain, capturing both strategic and tactical linkages between the two.

This section explores three related aspects that have emerged in the literature. First, empirical research faces several practical barriers and methodological limitations that have hampered widespread development. Next, research can be characterized using two major themes. One theme has explored how the two major streams of a CLSC, namely, the forward and reverse flows, can be strategically linked in terms of business models and competitive advantage. The second theme explores the environmental management implications of a CLSC.

13.2.1 Challenges for Empirical Research

Much of the research interest in CLSCs has taken a quantitative modeling perspective. Correspondingly, studies that have used empirical methodology (e.g., data and information from cases and surveys) are rather sparse (Kocabasoglu et al. 2007;

Prahinski and Kocabasoglu 2006). For example, a recent literature review of studies related to reverse logistics revealed that only about one-quarter were empirical, with a meager 5 percent employing larger datasets drawn from such sources as surveys* (Rubio et al. 2008). Moreover, much of the case-based research used a single case study to illustrate a conceptual model or provide parameter estimates for an analytic model.

Three practical barriers—that are slowly disappearing—explain the scarcity of empirical research, particularly from using large-scale surveys and archival data. First, there is a wide variety of operating contexts associated with CLSC, making the application of large-scale survey methods challenging. For instance, the European automotive industry has adopted a variety of closed-loop strategies that range from outsourcing the logistical management of product recovery (Krikke et al. 2004) to in-house engine remanufacturing facilities, to broad-scale integration across a firm's network of operations (Seitz and Peattie 2004; Seitz and Wells 2006).

A second barrier is the fact that CLSCs have only started to gain momentum over the last decade in several industries (Seitz and Wells 2006), limiting the potential pool of respondents that might be targeted. This limitation is likely to be less problematic in the near future, as many industries producing discrete goods have become the target of recent environmental laws and regulations. This growing governmental and public concern is being translated into more CLSC practices in a greater number of industries (French and LaForge 2006).

Finally, it remains unclear what is the best unit of analysis: plant, firm, or supply chain? Each has its merits and shortcomings. For example, as the degree of precision and target pool increases at the plant level, the scope of supply chain linkages, decision making, and analysis decreases. At the other extreme, conducting archival or survey research on multiple links in a supply chain is possible only in very unique circumstances.

Beyond these barriers, the characteristics of operating contexts for CLSC are highly heterogeneous across industries, and even within a specific industry, based on geography, regulations, and market segmentation. For example, Gonzalez-Torre et al. (2004) found systematic differences in the design of reverse logistics systems for packaging (i.e., bottles) even within the same regulatory regime driven by market differences. Product characteristics and the types of materials also matter greatly. Thus, CLSC activities can be further divided into those targeting the core product (Krikke et al. 2004), peripheral materials, such as packaging (Matthews 2004), and by-products from manufacturing, sometimes termed industrial ecology (Ehrenfeld and Gertler 1997). Moreover, products and materials can retain much

---

* The literature review covered 186 articles published between 1995 and 2005 in 26 academic journals including *California Management Review, European Journal of Operational Research, Harvard Business Review, Journal of Operations Management, Management Sciences, Operations Research,* and *Production and Operations Management.*

of their original intrinsic value (e.g., with remanufacturing) or very little (e.g., with recycling or energy from waste) of it.*

Depending on the targeted application, a firm's organizational structures and strategies also might differ. Formalized systems, despite the variety and complexity of products flowing through a reverse supply chain, are a critical factor for success (Autry 2005). Kocabasoglu et al. (2007), in an empirical study using survey data from the Canadian manufacturing sector, found a link between forward supply chain investments and low-value recovery (i.e., waste management/recycling), but not high-value recovery (i.e., reconditioning). Thus, low-value recovery may be viewed as a natural extension of the forward supply chain, but any management decision to develop or support high-value recovery CLSCs is driven by other strategic or tactical factors (see Chapter 2 for a discussion of some strategic issues involving CLSCs and Chapters 5 through 7 for a discussion of tactical issues). These results also suggest that the view managers might adopt when developing CLSCs will vary depending on the potential value to be reclaimed from core versus peripheral products. Moreover, for early-life returns such as catalog retailing, managers focus on value reclamation rather than any environmental benefits (Autry et al. 2001).

13.2.2 Strategic Linkages

Empirical research continues to develop and expand multiple conceptual models (and related research propositions) that attempt to describe characteristics and generalize outcomes from CLSCs (Prahinski and Kocabasoglu 2006; Rossi et al. 2006; Toffel 2004). For example, a firm's approach toward remanufacturing, one stage of the reverse supply chain, and its manufacturing strategy, which includes the forward supply chain, were compared in Guide et al. (2003) using the product–process matrix as a framework (Hayes and Wheelwright 1979). Based on three cases (i.e., Kodak, Xerox, U.S. Navy), several areas of alignment between the remanufacturing and manufacturing strategies were clear, including the positioning on the matrix with regard to volume, demand predictability, and complexity.

The timing of developing a CLSC can be viewed in a somewhat analogous fashion to other strategic capabilities: are there benefits to being a first mover, fast follower, or late adopter? At least for the automotive aftermarket industry, Richey et al. (2004) uncovered evidence that being a first mover into a market with a reverse logistics system was only beneficial if substantive resources were put in place; otherwise, being a fast follower provided more consistent benefits. Resources included technical, managerial, and financial components; managerial resources

---

* Some organizations also work to close the loop of the scrap material or unused inputs by returning them to the suppliers, redirecting them to another industry, or shipping them to a sister plant. For example, the scrap wood (wood chips, scrap log, etc.) from sawmill operations are redirected or sold to the pulp and paper plants (fibers) or to an energy-intensive industry for biomass.

were particularly important for providing more innovative, flexible reverse logistics systems. Finally, being a late adopter clearly hurts performance.

In their recent review article, Guide and Van Wassenhove (2009) synthesized a framework with three process-oriented components: (1) the front end, which is related to the concept of product acquisition (Guide and Van Wassenhove 2002); (2) the engine, which is associated with the actions of bringing back, testing, and reconditioning a product (Prahinski and Kocabasoglu 2006); and (3) the back end, which reintroduces the reconditioned product into the marketplace, either into the original market segment or another (Guide 2000). The framework provides a useful structure for developing analytic models and assessing much of the work to date; however, the potential importance of the forward supply chain in managing the entire closed loop is less clear. This may be partly due to semantics, as not all researchers define CLSC as the combination of the forward and reverse supply chains (French and LaForge 2006; Krikke et al. 2004). In contrast, empirical studies and conceptual/theoretical development tend to view this as foundational, and have proposed a clear strategic link (de la Fuente et al. 2008; Prahinski and Kocabasoglu 2006; Vachon et al. 2001).

One particular study combined the concepts of a responsive supply chain (Fisher 1997) and industry clockspeed (Fine 1998) to propose an optimal course of action in designing the reverse supply chain. More specifically, the length of a product life cycle and the design of a reverse supply chain is related (Blackburn et al. 2004). To identify the best fit, the marginal value of time must be assessed, which is inversely correlated to the duration of the product life cycle (hence positively related to the clockspeed). As the lifespan of parts and components from a product decrease, a "responsive" reverse supply chain should be used (in the same spirit of the Fisher's model), which must be capable of fast-tracking used products through the loop. In contrast, a returned product that is insensitive to time (i.e., low marginal value of time) is better serviced with an efficient reverse supply chain.

Interestingly, the forward supply chain, particularly during the design phase, can influence the effectiveness and efficiency of the reverse supply chain. The product design can include specifications that assist subsequent efforts to "fast track" parts in a reverse supply chain process or improve the efficiency of doing so. For example, the degree of standardization in products is an important variable in the marginal value of time. If a new product generation used a high number of standard parts/components, the marginal value of time will diminish.

Other critical features for the reverse supply chain are capabilities related to designing, and then building, products with a high degree of modularity (Krikke et al. 2004). Their study of the copier industry identified modularity as an excellent way to create value in an integrated forward–reverse supply chain. In fact, they proposed that product design should consider not only aspects of disassembly but also issues of repair, recycling, source reduction, and parts standardization. Ultimately, the reverse supply chain effectiveness is a function of the forward supply chain

(Vachon et al. 2001). In other words, a cradle-to-cradle approach (Rossi et al. 2006) can be foundational for improving the effectiveness of the CLSC.

It is also noteworthy that this relationship works in reverse as well: the reverse supply chain activities can lead to improvements in the forward supply chain. For example, a quick review of the dismantling operations at Frigidaire led to product design modifications (e.g., standardization of plastics parts and reduction of the number of parts in the product). These changes translated into a decrease in assembly time (by 76 percent) and a reduction of the space needed to perform the assembly of the refrigerators (Davis et al. 1997). HP offers another example; the information collected from their dismantling operations was fed back to product designers, which in turn increased the recyclability of newer generations of computers (Bartholomew 2002).

13.2.3 Environmental Management Implications

CLSC research often mentions environmental laws and regulations as a major driver for manufacturing organizations to develop a CLSC. The most widely cited are the European Community's waste of electrical and electronics equipment (WEEE) and end-of-life vehicle (EVL) (Kumar and Putnam 2008; Ostlin et al. 2008), and a series of take-back or "extended producer responsibility" regulations at the state and regional level in the United States (Toffel 2004). Also, the possibility for organizations to be associated with a "green" image has been cited as another environment-related driver for adopting a CLSC (Toffel 2004).

Despite this obvious recognition of environmental regulations, relatively little work has explored the systemic links with, and implications of, integrating environmental management into CLSC management. As a starting point, the operations strategy literature has proposed two extreme archetypes for considering external environmental pressures: constraint and integrated (Angell and Klassen 1999). The constraint perspective considers environmental management criteria as yet another product or service constraint, against which operations need to be buffered, or alternatively, flexible to accommodate. In doing so, the basic underlying operations process for the product or service changes little. To date, much of the CLSC literature has adopted a constraint approach where manufacturing organizations are limited in their structural and infrastructural elements when crafting their operations strategy.

In contrast, the integrated perspective requires a fundamental rethinking, which offers an extended view from which to redesign operational systems. Both environmental strategy and any CLSC design emerge and evolve through a series of feedback loops between a manufacturing organization, the external stakeholders, and competitive and regulatory forces. Gains are possible by expanding the boundaries of analysis and decision making (Corbett and Klassen 2006). Moreover, Daugherty et al. (2005) offer evidence of performance benefits from a relationship-oriented perspective. Thus, an explicit recognition is needed that environmental management decisions can be integrated into the overall supply chain systems, which in turn provides a greater scope of managerial discretion to the decision maker.

13.3 Linking Environmental Management and Closed-Loop Supply Chain

The claim that manufacturing organizations can benefit from CLSCs is essentially based on two premises. First, CLSCs generate value by either reducing costs or improving revenues in an era of increased competitiveness (Crandall 2006; Prahinski and Kocabasoglu 2006; Toffel 2004). Case studies and limited survey data provide some supporting evidence that customer satisfaction improves and costs are reduced from efforts to develop a CLSC (Autry 2005; Daugherty et al. 2001; Mollenkopf and Closs 2005; Richey et al. 2005).

Second, a CLSC can address environmental concerns voiced by a growing number of different stakeholder groups for greater environmental stewardship, such as customers, consumers, and the general population. However, as described in the previous section, the linkage between environmental management and CLSCs remains largely unexplored. To fill that gap, this section proposes a model that links the management of the environment and CLSCs. In particular, the model pulls together two main dimensions: environmental management orientation and CLSC integration.

13.3.1 Environmental Management Orientation

Environmental management has been the subject of numerous studies over the past 20 years, and like CLSC research, scholars have developed a wide array of definitions and operationalizations. However, a common typology is to consider a firm's positioning on its environmental strategy, which ranges on a spectrum going from a reactive behavior to a proactive behavior (Henriques and Sadorsky 1999; Klassen and Johnson 2004) (Figure 13.1).

A reactive orientation is compliance driven, and mainly aims to meet the legal requirements (Buysse and Verbeke 2002). Managers of these organizations have

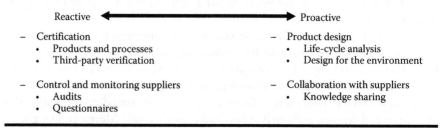

Figure 13.1 Environmental management orientation.

difficulty envisioning how environmental management might be a positive contributor to performance (Sharma and Vredenburg 1998). Also, management practices are characterized by minimizing environmental involvement, and reducing risks related to liabilities, fines, and accidents (e.g., spills). Risk minimization also can be extended to supply chain transactions using monitoring and control of suppliers' processes and components (e.g., using questionnaires and audits) (Vachon and Klassen 2006a). Thus, even a purchasing requirement to which suppliers provide an environmental certification (e.g., eco-label, ISO 14001) can be indicative of a reactive orientation, as it reduces the potential environmental risks (Klassen and Johnson 2004).

A proactive orientation is marked by product and process innovation that is not necessarily driven by external regulatory pressure (Sharma and Vredenburg 1998). In fact, environmental management initiatives can facilitate the development of broader organizational capabilities that provide other competitive benefits (Christmann 2000; Hart 1995; Russo and Fouts 1997). Particular environmental practices include design capabilities to incorporate new environmental features (e.g., reduced use of hazardous materials, which can simplify a CLSC), the application of life-cycle analysis when setting product specifications, and source reduction through process or product modifications (Klassen and Johnson 2004). Chapter 4 provides a summary of product design issues in CLSCs. Again, these initiatives frequently benefit from involving multiple supply chain partners in a collaborative fashion with the cross-fertilization of knowledge (Vachon and Klassen 2006a).

13.3.2 Closed-Loop Integration

As discussed earlier in this chapter, a CLSC can take several forms. For instance, the closed loop can be fully integrated into the overall organization. A classic example is Xerox's asset management program (Krut and Karasin 1999), where Xerox was involved in the entire reverse supply chain of their copiers. Moreover, the firm's competitive strategy and business model were predicated on integrating the forward and the reverse supply chains that leveraged financial lease agreements and component reuse (Vachon et al. 2001).

Another example is Caterpillar. Caterpillar performs remanufacturing of pumps, engines, and heavy equipment in several factories around the world (Brat 2006; Oster 2006). Remanufacturing is part of Caterpillar's strategy, and it is integrated throughout its plant network. However, this is not common practice in heavy equipment industry: most of Caterpillar's competitors have outsourced these activities to smaller regional organizations (Padley 2004). Thus, competitors are much less likely to capture related knowledge that subsequently can be leveraged in design, assembly, use, disassembly, and remanufacturing. Therefore, closed-loop activities can be undertaken with varying degrees of integration by the original

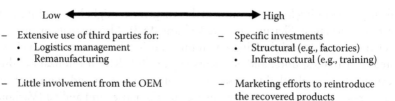

Figure 13.2   Closed loop integration.

equipment manufacturers (OEM). Figure 13.2 represents this array of integration of the CLSC within the existing supply chain.

Like Caterpillar's competitors, firms in many industries have little involvement in the regulations or other efforts to close the loop. For instance, auto recycling Netherland (ARN) is a joint venture among several car importers in the Netherlands, and coordinates recycling efforts for end-of-life vehicles. The use of a joint venture or industry-level association implies little integration of the closed loop in the supply chain of the OEM. Similarly, in the electronics industry, LG entered into a partnership with waste management to create a network of collection stations to recover e-waste (Makower et al. 2009). In other industries, the closed-loop integration is limited to the relatively straightforward use of recycled inputs. This practice is the only practical option in commodity-based industries such as paper and steel (Vachon et al. 2001).

Unfortunately, the thinking behind such a CLSC system design is very similar to that which drove many manufacturers to adopt end-of-pipe pollution controls when environmental issues first surfaced two or three decades ago. In the short term, an integrated closed loop is not inexpensive or straightforward. Investments are essential in facilities, labor training, and additional marketing efforts to reintroduce the reconditioned product on the market, but economies of scale and network effects are likely to lower long-term costs once sufficient infrastructure is in place. Finally, a high degree of integration implies increases in not only resource investments (Richey et al. 2005) but also operational risk. Hence, a low level of integration, as evidenced by outsourcing, denotes efforts to transfer operational risk elsewhere in the business model and supply chain.

13.3.3 Strategic Fit and Performance

Combining the two dimensions—environmental management orientation and closed-loop integration—offers the possibility to delineate different competitive options (Figure 13.3). Proactive environmental management is associated with the development of product design capabilities that employ design-for-the-environment principles and life-cycle analysis. Improved product design and enhanced

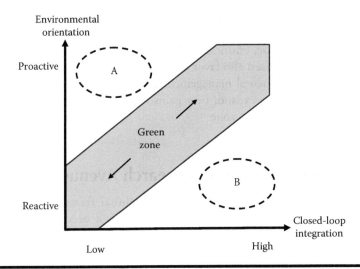

Figure 13.3 Linking environmental management orientation and CLSC integration.

parts/components specifications set the foundation for a greater recovery value from used products (e.g., durable parts, reusable parts), and a more efficient recovery process (i.e., ease of disassembly). As a result, an environmentally proactive firm is expected to use an integrated closed loop as the best option to secure recovery of the increased value of its used products, for subsequent reintroduction into the forward supply chain.

In contrast, a firm that has few, or very limited, environmental management capabilities is more likely to leave out considerations related to disassembly, repair, or reuse in the design of their products. Therefore, the same organization has less incentive to recover their products, as the net recovery value is relatively lower (given higher processing costs and lower recovery yields). Furthermore, it is unlikely that the same organization would invest resources in fully integrating the CLSC. Therefore, organizations that have a reactive environmental orientation stance will adopt a low level of closed-loop integration. Combined, the last two extremes set the endpoints for the diagonal region of Figure 13.3—the "green zone."

Note that Zones A and B in Figure 13.3 are less likely to occur. Zone A includes firms that have developed environmental management capabilities, but their products are of such a short life span, or entirely consumable, that investing in an integrated closed loop offers little competitive value. Packaging in the food industry is one such scenario. Industrywide design standards for packaging, however, would still facilitate greater economies of scale (i.e., lower cost) from a shared CLSC, possibly run by an independent third party, while acknowledging public pressure to minimize waste.

Zone B is also possible if particular components or materials, such as precious metals, are present in sufficient quantities to prompt the development of a vertically integrated CLSC. Likewise, business models that historically have been driven by leases (rather than purchases) also favor CLSCs, despite little concern for the environment or few environmental management capabilities. Yet, as Xerox found, as environmental capabilities expand, new gains and insights are possible, thereby moving the firm into the green zone.

13.4 Synthesis and Future Research Avenues

From the existing empirical research and conceptual framework synthesized here, several research avenues emerge. First, there is a need to further assess and operationalize the two dimensions (i.e., environmental management orientation and closed-loop integration) of the model with empirical evidence. Such a confirmation requires that the dimensions have discriminatory power in classifying the organizations. Although the literature has extensively addressed the measurement of environmental management orientation (Aragon-Correa 1998; Henriques and Sadorsky 1999; Russo and Fouts 1997; Sharma and Vredenburg 1998), much less has been done for closed-loop integration. Prior empirical research in forward supply chains represent a good starting point for measuring closed-loop integration (Ellinger et al. 2000; Frohlich and Westbrook 2001; Narasimhan 2001).

Once clear measures of integration have been validated, it is possible to go to the next step and assess the competitive value of being on or off the "green" diagonal, relative to either Zone A or Zone B. Attempting to do so, however, prompts a further empirical challenge: how is performance best assessed for a CLSC? What must be added or changed from classical forward supply chain performance metrics, such as cost, timeliness, and service quality, for CLSCs (Klassen 2009)? For instance, green supply chain management has been linked to traditional metrics such as financial performance (Zhu and Sarkis 2004) and operational performance (Vachon and Klassen 2008). However, little has been done for CLSCs, although initial efforts to assess customer satisfaction and cost containment for reverse logistics systems (Daugherty et al. 2001; Richey et al. 2005) have taken important tentative steps in that direction. All these research avenues confirm that CLSC remains a fertile topic for empirical research.

In this chapter, the literature review suggests that empirical research in CLSC is sparse and unfocused. In response, a conceptual framework that brings together two important strategic aspects—environmental orientation and closed-loop integration—is developed. The framework helps to delineate different competitive strategies with regard to CLSC and provides the basis for more structured and insightful empirical research.

References

Angell, L. C. and R. D. Klassen (1999), Integrating environmental issues into the mainstream: An agenda for research in operations management, *Journal of Operations Management*, 17 (3), 575–598.

Aragon-Correa, J. A. (1998), Strategic proactivity and firm approach to the natural environment, *Academy of Management Journal*, 41 (5), 556–567.

Autry, C. W. (2005), Formalization of reverse logistics programs: A strategy for managing liberalized returns, *Industrial Marketing Management*, 34 (7), 749–757.

Autry, C. W., P. J. Daugherty, and R. J. Richey (2001), The challenge of reverse logistics in catalog retailing, *International Journal of Physical Distribution & Logistics Management*, 31 (1), 26–37.

Bartholomew, D. (2002), Beyond the grave, *Industry Week*, 251 (3), 34–39.

Blackburn, J. D., V. D. R. Guide, G. C. Souza, and L. N. Van Wassenhove (2004), Reverse supply chains for commercial returns, *California Management Review*, 46 (2), 7–22.

Brat, I. (2006), Caterpillar gets bugs out of old equipment; Growing remanufacturing division is central to earnings-stabilization plan, *Wall Street Journal*, New York, July 5, A.16.

Buysse, K. and A. Verbeke (2002), Proactive environmental strategies: A stakeholder management perspective, *Strategic Management Journal*, 24 (5), 453–470.

Christmann, P. (2000), Effect of 'best practices' of environmental management on cost advantage: The role of complementary assets, *Academy of Management Journal*, 43 (4), 663–680.

Corbett, C. J. and R. D. Klassen (2006), Extending the horizons: Environmental excellence as key to improving operations, *Manufacturing & Service Operations Management*, 8 (1), 5–22.

Crandall, R. E. (2006), How green are your supply chains, *Industrial Management*, 48 (3), 6–11.

Daugherty, P. J., C. W. Autry, and A. E. Ellinger (2001), Reverse logistics: The relationship between resource commitment and program performance, *Journal of Business Logistics*, 22 (1), 107–123.

Daugherty, P. J., R. G. Richey, B. J. Hudgens, and C. W. Autry (2005), Reverse logistics in the automobile aftermarket industry, *International Journal of Logistics Management*, 14 (1), 49–61.

Davis, G., C. A. Wilt, P. S. Dillon, and B. K. Fishbein (1997), *Extended Product Responsibility: A New Principle for Product-Oriented Pollution Prevention*, Washington, DC: United States Environmental Protection Agency.

de la Fuente, M. V., L. Ros, and M. Cardos (2008), Integrating forward and reverse supply chains: Application to a metal-mechanic company, *International Journal of Production Economics*, 111 (2), 782–792.

Delmas, M. and M. W. Toffel (2008), Organizational responses to environmental demands: Opening the black box, *Strategic Management Journal*, 29 (10), 1027–1055.

Ehrenfeld, L. and N. Gertler (1997), Industrial ecology in practice: The evolution of interdependence at Kalundborg, *Journal of Industrial Ecology*, 1 (1), 67–79.

Ellinger, A. E., P. J. Daugherty, and S. B. Keller (2000), The relationship between marketing/ logistics interdepartmental integration and performance in U.S. manufacturing firms: An empirical study, *Journal of Business Logistics*, 21 (1), 1–22.

Fine, C. H. (1998), *Clockspeed: Winning Industry Control in the Age of Temporary Advantage*, Reading, MA: Perseus Books.

Fisher, M. L (1997), What is the right supply chain for your product? *Harvard Business Review*, 75 (2), 105–116.

French, M. L. and R. L. LaForge (2006), Closed-loop supply chains in process industries: An empirical study of producer re-use issues, *Journal of Operations Management*, 24 (3), 271–286.

Frohlich, M. T. and R. Westbrook (2001), Arcs of integration: An international study of supply chain strategies, *Journal of Operations Management*, 19 (2), 185–200.

Gonzalez-Torre, P. L., B. Adenso-Diaz, and H. Artiba (2004), Environmental and reverse logistics policies in European bottling and packaging firms, *International Journal of Production Economics*, 88 (1), 95–104.

Guide Jr., V. D. R. (2000), Production planning and control for remanufacturing: Industry practice and research needs, *Journal of Operations Management*, 18 (4), 467–483.

Guide, V. D. R. and L. N. Van Wassenhove (February 2002), The reverse supply chain, *Harvard Business Review*, 80 (2), 25–26.

Guide, V. D. R. and L. N. Van Wassenhove (2009), The evolution of closed-loop supply chain research, *Operations Research*, 57 (1), 10–18.

Guide, V. D. R., V. Jayaraman, R. Srivastava, and W. C. Benton (2000), Supply-chain management for recoverable manufacturing systems, *Interfaces*, 30 (3), 125–142.

Guide, V. D. R., V. Jayaraman, and J. D. Linton (2003), Building contingency planning for closed-loop supply chains with product recovery, *Journal of Operations Management*, 21 (3), 259–279.

Hart, S. L. (1995), A natural-resource-based view of the firm, *Academy of Management Review*, 20 (4), 986–1014.

Hayes, R. H. and S. C. Wheelwright (1979), Link manufacturing process and product life cycles, *Harvard of Business Review*, 57 (1), 133–140.

Henriques, I. and P. Sadorsky (1999), The relationship between environmental commitment and managerial perceptions of stakeholder importance, *Academy of Management Journal*, 42 (1), 87–99.

King, A. and M. Lenox (2002), Exploring the locus of profitable pollution reduction, *Management Science*, 48 (2), 289–299.

Klassen, R. D. (2009), Comment on The Evolution of Closed-Loop Supply Chain Research, V. D. R. Guide and L. N. Van Wassenhove, Operations Research (online commentary), 57 (1), http://orforum.blog.informs.org/files/2009/03/klassen.pdf

Klassen, R. D. and P. F. Johnson (2004), The green supply chain, in *Understanding Supply Chains: Concepts, Critiques and Futures*, S. New and R. Westbrook (Eds.), New York: Oxford University Press.

Klassen, R. D. and C. McLaughlin (1996), The impact of environmental management on firm performance, *Management Science*, 42 (8), 1199–1214.

Kocabasoglu, C., C. Prahinski, and R. D. Klassen (2007), Linking forward and reverse supply chain investments: The role of business uncertainty, *Journal of Operations Management*, 25 (6), 1141–1160.

Krikke, H., I. Le Blanc, and S. van de Velde (2004), Product modularity and the design of closed-loop supply chain, *California Management Review*, 46 (2), 23–39.

Krut, R. and L. Karasin (1999), Supply Chain Environmental Management: Lessons from Leaders in the Electronics Industry, United States-Asia Environmental Management.

Kumar, S. and V. Putnam (2008), Cradle-to-cradle: Reverse logistics strategies and opportunities across three industry sectors, *International Journal of Production Economics*, 115 (2), 305–315.

Makower, J., M. Wheeland, and T. Herrera (2009), *State of Green Business 2009*, Oakland, CA: Greener World Media, available at www.greenbiz.com

Matthews, H.S. (2004), Thinking outside 'the box': Designing a packaging take-back system, *California Management Review*, 46 (2), 105–119.

Mollenkopf, D. A. and D. J. Closs (2005), The hidden value in reverse logistics, *Supply Chain Management Review*, 9 (5), 34–43.

Narasimhan, R. (2001), The impact of purchasing integration and practices on manufacturing peformance, *Journal of Operations Management*, 19, 593–609.

Oster, S. (2006), Caterpillar, China are to promote remanufacturing, *Wall Street Journal*, New York, September 15, A.10.

Ostlin, J., E. Sundin, and M. Bjorkman (2008), Importance of closed-loop supply chain relationships for product remanufacturing, *International Journal of Production Economics*, 115 (2), 336–348.

Padley, K. (2004), Caterpillar expands remanufacturing operations, *Reuters*, December 23.

Prahinski, C. and C. Kocabasoglu (2006), Empirical research opportunities in reverse supply chain, *Omega*, 34 (6), 519–532.

Richey, R. G., P. J. Daugherty, S. E. Genchev, and C. W. Autry (2004), Reverse logistics: The impact of timing and resources, *Journal of Business Logistics*, 25 (2), 229–250.

Richey, R. G., S. E. Genchev, and P. J. Daugherty (2005), The role of resource commitment and innovation in reverse logistics performance, *International Journal of Physical Distribution & Logistics Management*, 35 (3/4), 233–257.

Rossi, M., S. Charon, G. Wing, and J. Ewell (2006), Design for the next generation: Incorporating cradle-to-cradle design into Herman Miller products, *Journal of Industrial Ecology*, 10 (4), 193–210.

Rubio, S., A. Chamoro, and F. J. Miranda (2008), Characteristics of the research on reverse logistics, *International Journal of Production Research*, 46 (4), 1099–1120.

Russo, M. V. and P. A. Fouts (1997), A resource-based perspective on corporate environmental performance and profitability, *Academy of Management Journal*, 40 (3), 534–559.

Seitz, M. A. and K. Peattie (2004), Meeting the closed-loop challenge: The case of remanufacturing, *California Management Review*, 46 (2), 74–89.

Seitz, M. A. and P. E. Wells (2006), Challenging the implementation of corporate sustainability: The case of automotive engine remanufacturing, *Business Process Management Journal*, 12 (6), 822–836.

Sharma, S. and I. Henriques (2005), Stakeholder influences on sustainability practices in the Canadian forest products industry, *Strategic Management Journal*, 26 (2), 159–180.

Sharma, S. and H. Vredenburg (1998), Proactive corporate environmental strategy and the development of competitively valuable organizational capabilities, *Strategic Management Journal*, 19 (8), 729–753.

Stock, J. R. (1998), *Development and Implementation of Reverse Logistics Programs*. Oak Brook, IL: Council of Logistics Management.

Toffel, M. W. (2004), Strategic management of product recovery, *California Management Review*, 46 (2), 120–141.

Vachon, S. and R. D. Klassen (2006a), Extending green practices across the supply chain: The impact of upstream and downstream integration, *International Journal of Operations and Production Management*, 26 (7), 795–821.

Vachon, S. and R. D. Klassen (2006b), Green project partnership in the supply chain: The case of the package printing industry, *Journal of Cleaner Production*, 14 (6–7), 661–671.

Vachon, S. and R. D. Klassen (2008), Environmental management and manufacturing performance: The role of collaboration in the supply chain, *International Journal of Production Economics*, 111 (2), 299–315.

Vachon, S., R. D. Klassen, and P. F. Johnson (2001), Customers as green suppliers: Managing the complexity of the reverse supply chain, in *Greening Manufacturing: From Design to Delivery and Back*, J. Sarkis (Ed.), Sheffield, U.K.: Greenleaf Publisher.

Zhu, Q. and J. Sarkis (2004), Relationships between operational practices and performance among early adopters of green supply chain management practices in Chinese manufacturing enterprises, *Journal of Operations Management*, 22 (3), 265–289.

Chapter 14

Conclusion and Future Research Directions

Mark Ferguson and Gilvan C. Souza

Content

Our objective in writing this book was to provide a concise and easy-to-read summary of the latest research in the field of closed-loop supply chains (CLSCs) to both practitioners and academics. We had contributions from some of the best-known experts in this growing field. Part I of this book, with three chapters, was devoted to strategic decisions facing firms regarding the secondary market for its products: whether to pursue remanufacturing, leasing versus selling, the impact of take-back legislation, and guidelines for product design for CLSCs. Part II, with four chapters, was devoted to more tactical issues, once the decision to "close the loop" has been made. Issues explored include network design (for collection, consolidation, and reprocessing centers), used-product acquisition, grading and disposition, production planning for remanufactured products, and marketing issues for remanufactured products—pricing, positioning, and warranties. Part III was devoted to case studies: one chapter was devoted to the motion picture industry, one chapter to eco-efficiency initiatives in hospitals, and one chapter to the discussion of practices in retreaded tires, computers and printers, printer cartridges, and IT networking equipment. Finally, Part IV discussed the interdisciplinary nature of CLSC research, including a chapter summarizing empirical findings, and another chapter summarizing case studies presented at a recent workshop, as well as how CLSC research has touched upon marketing, product design, product-service systems, industrial ecology, and the economic development of distressed communities.

The enormous growth in CLSC research over the past ten years, as evidenced by the materials summarized in this book, is a response to the growing societal and business emphasis on the triple bottom line of sustainability. It is refreshing to see the academic community providing useful insights to practitioners. As already noted by Guide and Van Wassenhove (2009), it is imperative that the academic community continues to be relevant by focusing on problems of clear practical value. Empirical research (see Chapter 13), broadly defined to include normative research based on actual case studies, should be emphasized. Normative modeling research, the bread and butter of the CLSC community, should continue to provide decision-support tools based on real problems faced by companies, in a way that can be generalized to a broader set of industries. For example, the methodology presented in Chapter 7 for production planning was built from the interactions with one company—Pitney Bowes—but it is general enough to be applicable to companies where products have a reasonably long life cycle; a considerable amount of returns originate from leasing (making it easier to predict the return stream), and remanufactured product demand can be forecasted.

Regarding specific topics of future research, we emphasize the following. First, we believe the interface with industrial ecology should be strengthened, as emphasized in Chapter 12. Given the increased focus on the triple bottom line, research should attempt to incorporate environmental and societal impacts to the extent possible, particularly in decision-support models that have traditionally focused on cost minimization or, more recently, on profit maximization, as indicated in Chapter 6. As discussed in that chapter, a significant portion of research on product acquisition and disposition has focused on cases where the firm minimizes the cost of meeting the demand for remanufactured products, although a recent application has taken a revenue management perspective to the problem, considering profit margins and demand uncertainty for disposition alternatives (such as dismantling for spare parts). The next step would logically incorporate the environmental impacts of different disposition alternatives into the objective function, a recognizably difficult task.

Second, we believe that design considerations have played a limited role into the decision-support models of the CLSC literature. Product design, however, is a key determinant of life-cycle costs, as evidenced by Xerox's Asset Recovery Management program, where the core of the product is designed to last for multiple generations, despite its increased cost due to more durable designs, and a recent teaching case of Herman Miller and its incorporation of the cradle-to-cradle design protocol for one of its high-end office chairs (Lee and Bony 2008). Chapter 4 provides an overall treatment of product design for CLSC management from an engineering perspective, offering several design guidelines. However, the community should attempt to model how the choice of different design alternatives impacts the firm's (triple) bottom line.

Finally, we agree with the call made for more empirical work, which has been previously mentioned by several authors. In particular, there is a need for publicly available secondary datasets, based on actual practice, where various models and hypothesis can be tested and benchmarked against each other. The data can be disguised in a way that it protects the confidentiality of the firm where the data was collected from, but still maintains the basic structure needed for estimating the parameter values of various models. Examples of such papers in other research areas include Bodea et al. (2009) and Willems (2008). Of course, data from a single firm may not be representative of the larger population of firms; so a categorization of the practices in different industries is also needed. We provided a few examples of this in Chapter 9, but a more extensive study is needed. Such a study should help reduce the "selective" justification of modeling assumptions from different industry examples that sometimes occurs in our field, where the assumptions are based more on modeling convenience rather than accurately capturing the key drivers of the problem being modeled. As with any research field, the better we understand the current practices and business environments the more relevant and useful our research will be.

References

Bodea, T., Ferguson, M., and L. Garrow. 2009. Choice-based revenue management: Data from a major hotel chain. *Manufacturing and Services Operations Management*, 11(2), 356–361.

Guide, D. and L. Van Wassenhove. 2009. OR FORUM—The evolution of closed-loop supply chain research. *Operations Research* 57(1), 10–18.

Lee, D. and L. Bony. 2008. Cradle-to-Cradle Design at Herman Miller: Moving Toward Environmental Sustainability. Harvard Business School Case 9-607-003.

Willems, S. P. 2008. Data set: Real-world multi-echelon supply chains used for inventory optimization. *Manufacturing & Service Operations Management*, 10(1), 19–23.

Index